Marie-Louise von Franz

Rhythm and repose

with 143 illustrations, 16 in colour

Thames and Hudson

ART AND IMAGINATION
General Editor: Jill Purce

Filmset in Great Britain by Keyspools Ltd, Golborne, Lancashire
Printed and bound in Spain

Contents

Time as a deity and the stream of events

Time is one of the great archetypal experiences of man, and has eluded all our attempts towards a completely rational explanation.[1] No wonder that it was originally looked upon as a Deity, even as a form of manifestation of the Supreme Deity, from which it flows like a river of life. Only in modern western physics has time become part of a mathematical framework, which we use with our conscious mind to describe physical events. The mind of primitive man made less distinction than ours between outer and inner, material and psychic, events. Primitive man lived in a stream of inner and outer experience which brought along a different cluster of coexisting events at every moment, and thus constantly changed, quantitatively and qualitatively.

Even our seemingly self-evident concepts of past, present and future do not seem to be universal. The Hopi Indians, for instance, do not possess them in their language. Their universe has two basic aspects: that which is manifest and thus more 'objective' and that which is beginning to manifest and is more 'subjective'. Concrete objects are manifest and in this way already belong to the past; inner images, representations, expectations and feelings are 'subjective', on their way to manifestation, and thus bend more towards the future. The present is that razor's-edge where something stops beginning to manifest (is already past) or is on the verge of beginning to manifest. There is no continuing flow of time for the Hopi, but a multiplicity of subtly distinguished events. The creator of all this is 'a'ne himu, a 'Powerful Something' which is a kind of cosmic breath.[2] Children, too, do not at once live in our communal clock time. They have been shown to perceive rhythm, velocity and frequency long before they begin to adapt to our ordinary notion of time.[3]

In man's original point of view time was life itself and its divine mystery. This remains so in the ancient Greek notion of time. The Greeks actually identified time with the divine river Oceanos, which surrounded the earth in a circle and which also encompassed the universe in the form of a circular stream or a tail-eating serpent with the Zodiac on its back. It was also called Chronos (Time) and later identified with Kronos, the father of Zeus, and also with the god Aion.

1 Cf. Fraser, *The Voices of Time* and *The Study of Time*. Cf. also Fraser, *Of Time, Passion and Knowledge*; Whitrow, *The Natural Philosophy of Time*; Priestley; Le Lionnais.

2 Cf. Whorf, 80.

3 Cf. Piaget.

Aion originally denoted the vital fluid in living beings, and thus their life-span and allotted fate. This fluid continued to exist after death in the form of a snake.[4] It was a 'generative substance', as was all water on earth and especially Oceanos-Chronos, the creator and destroyer of everything. The philosopher Pherekydes taught that the basic substance of the universe was time (Chronos), from which fire, air and water were produced.[5] Oceanos was also a kind of primal World Soul.[6]

In Hellenistic times, Aion Chronos was identified with the old Persian time-god Zurvān.[7] The ancient Persians discerned two aspects of this supreme deity: Zurvān akarana, Infinite Time, and Zurvān dareghō-chvadhātā, Time of a Long Dominion. The latter was the cause of decay and death, and was sometimes even identified with Ahriman, the principle of evil. The Orphic and Mithraic circles of late antiquity identified Zurvān, in both his opposed aspects at once, with their Aion.[8] A text invokes Aion with the following words:

> I greet thee, thou that fillest the whole structure of the air, spirit that stretchest from heaven to earth . . . and to the confines of the abyss . . . spirit that also penetratest myself and leavest me again. Thou, the servant of the rays of the sun, that enlightenest the world . . . a great circular mysterious form of the universe, heavenly spirit, ethereal spirit, earthy, fiery, windy, light . . . dark spirit, that shinest like a star. . . . Lord, god of the Aions. . . . Ruler of everything.[9]

Aion, the god of time, is here clearly an image of the dynamic aspect of existence, of what we might call today a principle of psycho-physical energy. All opposites – change and duration, even good and evil, life and death – are included in this cosmic principle.

This Aion was also sometimes identified with the sun god, who is obviously the great indicator of time measures. The initiate prays to him: 'O Lord, who with thy spirit bindest the fiery keys of the fourfold belt . . . fire-walker, creator of light, fire-breather, with fiery courage . . . Aion, lord of light, . . . open the doors to me.'[10]

This god inherited many aspects of the Egyptian sun god, Ra, who was the ruler of time. Every hour of the day and night this supreme deity changed his shape: he rose, for instance, as a scarab, and descended into the underworld as a crocodile; in the moment of his resurrection after midnight he assumed the shape of a double lion, Routi, 'Yesterday and Tomorrow'.[11] Osiris, the resurrected human being who became a god, also says of himself: 'I am Yesterday, Today and Tomorrow.'[12] He lives in the 'House of Eternity' or a 'House of Millions of Years'.

Besides the sun god, the ancient Egyptians personified unending time also as a separate god, Ḥeḥ, who has the Ankh, the symbol of life, suspended at his right arm. As in Greece, the snake in Egypt was also connected with time. It symbolized life and health, and each individual was protected by a 'life-time snake', which was a daemon of time and of survival after death.[13]

This same archetypal symbolism of time, as the godhead and also as an unending stream of life and death, can be found in India. In the *Bhagavadgīta* (3rd or 4th century BC) the god Krishna reveals himself to Arjuna in his terrible form. He sees in him all other gods together: 'I see a figure infinite,

4 Cf. Onians, 206.

5 *Ibid.*, 248.

6 *Ibid.*, 249.

7 Cf. Brandon *History, Time and Deity* and 'The Deification of Time', 376.

8 Cf. Campbell.

9 Preisendanz, I, 111.

10 *Ibid.*, 93.

11 Cf. De Wit, 72; and von Franz, *Number and Time*, 92ff.

12 Cf. Brandon, 'The Deification of Time', 372.

13 Cf. Bonnet, 682, 257ff, 833–34.

wherein all figures blend to countless bodies, arms and eyes.' In a great stream they disappear into his jaws of flaming fire, entering them with hurried step.[14] Then Vishnu says: 'Know I am Time that makes the worlds perish, when ripe and come to bring them destruction.' Not only Vishnu but also Shiva represents time. He symbolizes 'the energy of the universe increasingly creating and sustaining the forms in which he manifests himself'.[15] Shiva is called Mahā Kāla, 'Great Time', or Kāla Rudra, 'all-devouring Time'. His Shakti, or active energy, appears in its destructive form personified in the terrible goddess Kālī, who is Time, for Kālī is the feminine form of Kāla, which means Time, the black-blue colour, and death.

The mystical philosophy of Hinduism looks upon this world as unreal; time is especially what deceives the unenlightened soul into believing exclusively in his own self-conscious being and the reality of outer things. But in fact this perishable, changing world is a kind of illusion: 'Verily for him who knows this (for the enlightened man) . . . the sun never rises nor sets. For him it is day forever' (*Chandogya Upanishad* III, 11.3).

This 'eternal day' is God himself. Similarly, a North American Delaware Indian described a vision he had of God: He saw 'a great man clothed with the day, the most radiant day he had ever seen, a day of many years, even of eternal duration. The whole world was spread over Him, so that one could see the earth and all things on earth in Him.'[16]

In China the supreme deity had not always been personified, but here too time is seen as an aspect of the dynamic, creative basic principle of the universe. Time thus belongs to the masculine Yang principle, which is symbolized by three straight lines; its female counterpart Yin (symbolized by three broken lines) is associated with space. These two together manifest the Tao, the secret law which governs the cosmos. Yang, the Creative, 'acts in the world of the invisible with Spirit and Time for its field, Yin the Receptive acts upon Matter in Space and brings material things to completion'.[17] Time, seen in this way, is 'the means of making actual what is potential'. The group of lines which symbolizes Yang when doubled is called Ch'ien (Heaven) and denotes the movement of heaven which is unending. This duration in time is the image of the power inherent in the creative principle; it is symbolized by the dragon. It produces quality, while the receptive produces quantity.[18]

As Marcel Granet has shown,[19] Yan and Yin are not static cosmic principles but alternating cosmic *rhythms*. Time has never been considered in China as an abstract parameter or as an 'empty' time period. The word for time, che, means rather a circumstance favourable or unfavourable for action. Time and space, says Granet, 'were considered as an *ensemble* [grouping, cluster] of occasions and places',[20] a bundle of coinciding events. What 'bundles' such occasions together is called chin, 'duration': 'Former times, the present time, the morning and the evening are combined together to form duration. . . . Time however sometimes has no duration, for the beginning point of time has no duration.'[21] As long as they are 'germs', situations are still outside time and can be influenced by men; only when they have entered duration-time do they become fixed entities.

A close relationship of time with the cosmic creative energy emanating from the godhead can also be found in the Maya and Aztec notions of time.

14 *The Bhagavadgita*, ch. 11.

15 Cf. Zimmer, 148–51.

16 Müller.

17 Cf. Wilhelm, I, 307.

18 *Ibid.*, 324, 367.

19 Granet, 79.

20 *Ibid.*,

21 Cf. Needham, *Time and Eastern Man*.

The most frequently used word for time in the Maya language, *kin*, is most often represented by the hieroglyph shown here, which means 'sun' and 'day'.[22] It consists of a four-petalled flower, a species of plumeria. The emanating lines below are the 'sun's beard' or 'cords of the sun' or 'arrowshafts of the sun'; in my opinion, they represent the creative vital energy of the sun. The snake is also not missing in this connection: the Maya worshipped a double-headed snake whose one head meant life, the other death.[23]

In the Aztec civilization time was associated with the supreme deity, with the creator-god Omotéotl, mother and father of all things, who was called 'mirror that illumines all things', Lord of Fire and Lord of Time. This god first created four other gods: the red Tezcatlipoca, placed in the east; the black one, who lives in the north; the white one in the west; and the blue one in the south. To these four gods belong certain plants, animals, and qualitatively different years. In the middle dwells the god of fire. The four Tetzcatlipocas, then, created all other things. Only with them did space and time fully enter the world. The idea of time contains in the Aztec idea something abrupt. At one time east and positive forces dominate, at another the north and austerity; today we live in good times, tomorrow perhaps in unfavourable days. 'Is there perchance any truth in our words here?' says a wise man and poet; 'all seems so like a dream.' Only after death, says another, shall we 'know His face', namely that of the supreme god whose name is 'Night and Wind', and 'who is an inscrutable mystery'.[24]

In contrast to these myths in which God Himself *is* time (and also not-time) itself, our own Judaeo-Christian tradition sees God as purely outside time, as having created time together with the universe. After God separated the waters above the firmament from the waters below, and created the sun and moon, day and night came into existence and time began. And though we believe that material nature obeys laws, which make certain events recur in time, there are also recurrent miracles, magical and parapsychological phenomena, which are caused by the direct intervention of the creator God, a constant dramatic confrontation with His creation and with man. The most radical of these events, which disrupted time into a completely different Before and After, is the incarnation of Christ. According to I Peter 3:18, Christ died but once for our sins, once and for all (*hapax, semel*). Thus the development of history is governed and oriented by a unique fact which can never be repeated. On account of Christ's promise to return, the early Christian congregations were oriented much more to the future than to the past, hoping for Christ's return in glory.[25] In a similar way the Jews expect the coming of the Messiah at the end of time.

With St Augustine a new aspect of this idea of time entered into our tradition: the idea that God is present not only in the cosmos but also in man's innermost soul. Thus time too, being a 'working' of God, acquires a psychological nuance. The present is nothing if not an experience in the soul; the past is a memory image in the soul; and the future exists only as our psychic expectations. But ordinary time is transient and meaningless: it disappears when the soul unites with God.[26]

Later I will discuss the partial return to a cyclic notion of time in Christian civilization; what never disappeared was the association of time with the

22 Cf. Morley, 72. For the picture, see Thompson, 142ff. I owe this information to the kindness of Dr med. José Zavala.

23 Cf. Krickeberg *et al.*, 65

24 León-Portilla, *Aztec Thought and Culture*, 73.

25 Cf. Quispel, 89.

26 *Ibid.*, 99–104.

idea of God's direct interference in the world. Even for Isaac Newton there still existed an absolute space and an absolute flux of time, which were both emanations from God, though in practical physics they had become parameters, time being measurable through moving bodies. This incipient separation of 'divine time' from measurable time is not unrelated to the development of the clock, our instrument of time measurement.[27]

The original image or intuition of time as a river or flow underlies those time-measuring devices which were based on the flow of some substance, a liquid – in water clocks and mercury clocks – or sand.

Either the Chaldaeans or the Egyptians seem to have first invented the clepsydra or water clock. Water flowed from an upper receptacle into a lower one which was graduated. (This measurement was based on an error, because in fact when the upper pressure diminishes the water flows more slowly.) In classical antiquity clepsydras were in widespread use. Around the year 100 BC there was one installed in the market in Athens to indicate time officially. In Athens and Rome the courts of justice used them to time and limit speeches. Water clocks were in use in Europe and Asia throughout the Middle Ages. The hourglass or sand clock, which is based on the same idea of time as a flux, was always less accurate. It seems to have been in use only since the fourteenth century.

The other symbol of cosmic energy, in addition to water, is fire, as we saw in the many divine images mentioned above. Thus fire too was used to measure time. The Arab Al-Yazari, in a treatise of AD 1206, describes a light clock, a candle which burned for thirteen hours. It contained little balls in holes on different levels. Each hour a ball fell down and activated a little mechanical figure, which trimmed the wick. It was the Chinese who made the most use of fire to measure time. This was done by spreading a combustible powder round a generally circular labyrinth and igniting it at one end, so that its burning crept slowly forward like a fuse. Conventional phrases, such as 'long life' or 'double good fortune', were often placed in the centre, and pebbles attached with a thread at certain places fell down when the thread burned up and were used, by their fall, to awaken the sleeper.

These systems which measure time by the flux of a substance or the burning of a powder imply that time is a linear flow, and similarly Western classical physics used a line to represent time, alongside the three parameters of Euclidian space, for all its measurements and descriptions of physical events. The great change came through Albert Einstein, who realized that temporal indications were always relative to the position of the observer. Only because the velocity of light is so high – 186,000 miles per second – can we ignore this in the practical macrophysical realm; as soon as the observer also moves with a high velocity, the time span between the event and its observation becomes a problem for establishing a sequence of events. Two events which are seen as occurring simultaneously by one observer may occur as different temporal sequences for others. In high-energy physics, where we observe interactions between nuclear particles that move almost at the speed of light, time is completely relative. The idea of the space-time coordinate system as something objective is no longer valid. It is only a tool used by an observer to describe his special environment.[28]

27 Cf. Le Lionnais, 22.
28 Cf. Capra, 105 and elsewhere.

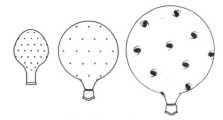

Block universe in three-dimensional approximation

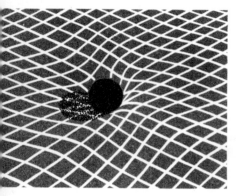

Mass produces a curvature in space

Einstein further realized that the requirement that the laws of nature be formulated in such a way that they have the same form in all coordinate systems, that is, for all observers in whatever positions and motions, can be satisfied in the description of electromagnetic phenomena only if all spatial and temporal specifications are relative. Every change of coordinate systems mixes space and time in a mathematically defined way. Space and time are thus inseparably connected and form a four-dimensional continuum – which is generally called the Minkowski-Einsteinian block universe. This insight of Einstein's appears to be a return, on a higher level and with mathematical precision, to the age-old primitive intuition whereby, for instance, the Aztec god Omotéotl, with the four Tetzcatlipocas in the four corners of space, created space and time simultaneously. Fritjof Capra quotes the *Avatamsaka Sutra* of Mahayana Buddhism,[29] which asserts that in a state of illumined dissolution we lose the distinction between mind and body, subject and object, and realize that every object is related to every other object, not only spatially but also temporally: 'As a fact of pure experience, there is no space without time, no time without space. They are interpenetrating.'

Einstein went one step further in his special theory of relativity, a step which in a strange way revives the primitive intuition of time as a flow of inner and outer events, now grasped in a precise mathematical formalism. This step means including gravity within the picture of space-time, for it makes space-time curved.[30] This is caused by the gravitational fields of massive bodies filling it. In a curved space-time the curvature affects not only the geometry of space but also the lengths of time intervals: 'Time does not flow at the same rate as in a "flat space-time" and as the curvature varies from place to place according to the distribution of massive bodies, so does the flow of time.'[31]

It is a remarkable coincidence that, at approximately the same time as physicists discovered the relativity of time in their field, C.G. Jung came across the same fact in his exploration of the human unconscious. In the world of dreams, time also appears as relative and the categories of 'before' and 'after' seem to lose their meaning. If we go as deep as the archetypal layer of the unconscious, time even seems to disappear completely. Man has always known this, in a way: all over the world stories are told in which a person goes into a fairy hill, into paradise, into the realm of death or into the kingdom of dwarfs, and when he returns, thinking that he has spent only one evening or night there, finds all his contemporaries dead; his village has vanished, and he hears that only a vague rumour has survived of a man having disappeared hundreds of years before. Washington Irving's *Rip van Winkle* is just one example of this type of story.

Whenever we touch the deeper archetypal reality of the psyche, it permeates us with a feeling of being in contact with something infinite. But, as Jung pointed out,[32] this is the telling question of our life: whether we are related to something infinite or not. 'Only if we know that what truly matters is the infinite can we avoid fixing our interests on futilities. In the final analysis we count for something only because of the essential we embody, and if we do not embody that life is wasted.'

The most exciting application of Einstein's new notion of space-time can

29 *Ibid.*, 98, 172n.

30 For this and the following, see Capra, 173ff.

31 *Ibid.*, 177.

32 Jung, *Memories, Dreams, Reflections*, ch. XI, 325ff.

33 Capra, 169.

be found in its application in astrophysics. As astronomers and astrophysicists have to deal with very large distances, even light takes a long time to travel from the observed object to the observer. Thus the astronomer never looks at an object in its present state, but in its past. With our telescopes we can see galaxies which in fact existed millions of years ago. We can look at stars and clusters of stars at all stages of their evolution, looking at them, so to speak, backwards in time.[33]

The same applies to the effects of gravity. Because stars and galaxies are extremely massive bodies, the curvature of space-time becomes a relevant phenomenon. Its most extreme effects become manifest during a so-called gravitational collapse of a massive body – as happens, so we assume today, in the black holes. Due to the mutual gravitational attraction of its particles, which increases as the distance between the particles decreases, the star becomes more and more dense and thus space-time more and more curved, until finally even light can no longer escape from its surface. Thus a so-called 'event horizon' forms around the star, beyond which nothing is any longer observable and no clock signals can reach us; – in a way, the star walks out of time for us. It seems to me that something analogous might happen to us in death. When C. G. Jung died, on 6 June 1961, a patient of mine who did not know him dreamed: There were many people on a sunny day on a meadow; Jung was among them. He wore a suit which was green in front, black on the back. There was a black wall with a hole cut out exactly matching the outlines of Jung. He stepped into it, and so one saw now only a black wall, but she knew that he was still there, although invisible. She looked at herself and saw that she wore an identical green-black frock. In dying we may only step outside the 'event horizon' of the living, but still exist in an unobservable state.

Cyclical and linear time

Two aspects which belong to the primordial archetypal idea of time have already been touched upon in our mythological examples: the irreversible linear character of time and its cyclical aspect. The latter, which seems to predominate in most primitive civilizations, is probably based on the observation of the regular motion of the heavenly luminaries, and of the recurring seasonal changes. The circular river Oceanos and the tail-eating Zodiac snake imply this idea. Chronos-Kronos was directly called the 'round element' and also the 'giver of measures'. Macrobius writes: 'Insofar as time is a fixed measure it is derived from the revolutions of the sky. Time begins there, and from this is believed to have been born Kronos who is Chronos. This Kronos-Saturn is the creator of time.'[1]

In India a completely cyclical notion of time was predominant. The primary unit of time was the yuga or age (1,080,000 years). A complete cycle, or mahāyuga, consists of four such yugas, the number four signifying totality or perfection.[2] The first yuga of each cycle is a kind of Golden Age; then each yuga is worse than the last until at the end comes the 'great dissolution', and then the process begins again. The names of the yugas are taken from throws of dice. One mahāyuga or great year consists of 12,000 'divine years', each comprising 360 ordinary years – a total of 4,320,000 years. Thousands of such mahāyugas constitute a kalpa ('form'), which is

1 Macrobius, 6–8.
2 Cf. Eliade, 'Time and Eternity in Indian Thought', 177ff.

equivalent to a day in the life of Brahma. In this way time consists of a cosmic rhythm, a periodic destruction and re-creation of the universe.

For man, this cyclical aspect of time, viewed negatively, gives rise to Samsara, the ever-rotating wheel of birth and death, of endless reincarnations. Only the enlightened yogi or Buddhist who has understood Brahman or the Buddha-Mind in himself 'in a lightning flash of truth' is delivered in this life and can escape rebirth.[3] He has transcended the play of opposites and arrested all memory processes for ever.[4]

Mircea Eliade has shown that in many other civilizations there exists a slightly different myth of Eternal Return. The idea is that at the (extratemporal) moment of creation (which he calls *illud tempus*, 'that time') the archetypal models of all things and all human actions on earth came into existence.[5] However, the earthly replicas of these archetypes show a tendency towards deterioration and decay. Through retelling the myths of creation and by re-enacting the original rituals, man can renew the archetypal patterns and restore his own life forms. Similarly, with us Easter is considered not only the festival of Christ's resurrection, but of a total renewal of creation.[6] In Christianity the idea of *illud tempus* has become partly internalized: paradise or the Kingdom of Heaven is within ourselves and can be reached at any time by metanoia: by a basic change of attitude.[7]

The Chinese, too, knew a primordial time (*illud tempus*) when the culture heroes set all the patterns of life. They had a cyclical time concept (along with a linear one, which will be discussed later). At the bottom of the Chinese idea of time, as it underlies the *I Ching, the Book of Changes*, there were two circular time-models or time-mandalas. One was the so-called Sequence of Earlier Heaven, or Primal Arrangement, a circle built by the eight Kuas, the basic principles of all existence. Yang, the Creative or Heaven (Ch'ien), was placed in the south; Yin, the Receptive or Earth (K'un), was in the north. The whole sequence was arranged as shown here. This system is in a way timeless, though not without motion.

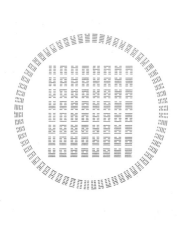

> 'Within the Primal Arrangement,' says Richard Wilhelm, 'the forces always take effect as pairs of opposites. Thunder . . . awakens the seeds of the old year. Its opposite, the wind, dissolves the rigidity of the winter ice. The rain moistens the seeds . . . while its opposite, the sun, provides the necessary warmth. Hence the saying: "Water and fire do not combat each other." . . . Keeping still stops further expansion. . . . Its opposite, the Joyous, bring about the joys of the harvest. Finally . . . the Creative, representing the great law of existence, and the Receptive, representing shelter in the womb, into which everything returns after completing the cycle of life.'[8]

Thus the opposites do not conflict: on the contrary they balance each other. This Primal Arrangement was associated with an arithmetical mandala called the Ho-t'u:

3 *Kausitaki Upanishad* IV, 2, quoted by Eliade in *Man and Time*, 187; and in *The Myth of the Eternal Return*, 113.

4 Eliade *The Myth of the Eternal Return*, 113.

5 *Ibid.*, 105.

6 *Ibid.*, 59.

7 *Ibid.*, 129.

8 Wilhelm, I, 287.

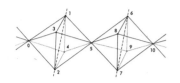

This movement repeats itself externally inside the arrangement.

According to Chinese tradition, King Wên and the Duke of Chou discovered, while they were imprisoned by the tyrant Chou Hsin (about 1150 BC), a new time mandala called the Later Heaven or Inner World Arrangement. In it Li (Fire) is in the south instead of Ch'ien (Heaven), K'an (Water) in the north instead of K'un (the Receptive). The trigrams are no longer grouped in pairs of opposites, but are shown in *a temporal progression* in which they manifest themselves in the phenomenal world through the cycle of years.

It shows 'God's activity in nature'. The Earlier Heaven emphasizes Duration, the Later emphasizes motion, but both are circular in form. The Later Heaven was also associated with a number mandala, the so-called Lo-shu (pattern of the river Lo), which was considered *the* basic numerical pattern of the universe.

$$
\begin{array}{ccc}
4 & 9 & 2 \\
3 & 5 & 7 \\
8 & 1 & 6
\end{array}
$$

This is a so-called magic square, in which all columns, rows and diagonals add up to 15. This square was regarded in China as a basic pattern of the universe, according to which architecture, music and even menus were arranged.[9]

Of course the Chinese had also an astrological system of time which resembles our own tradition.[10] However, its solar zodiac contains different animals and figures from ours; and our signs are monthly, whereas the Chinese are annual symbols.

Close resemblances to the astrological systems of China are visible in the Mayan and Aztec calendars. Not only did the Maya consider time as a deity (the sun god) but every year, month, day and even hour was identical with a number and was at the same time a god. The same holds true for the Aztecs, who also had such a time mandala. Though time existed potentially from the beginning as a principle connected with the supreme Lord of Time, it actually became manifest only after the creation of the four Tezcatlipocas. After a period of equilibrium, which resembles the balance of opposites in the Chinese Sequence of Earlier Heaven, each of these Tezcatlipocas wanted to become the sun. Thus strife and change came into existence. This led to a linear view of time which unfolds in five successive aeons or Suns. The Sun '4 Tiger' lasted 676 years; then the people were eaten by ocelots and the sun destroyed. Then followed the Sun called '4 Wind', which ended with everything being carried away by the wind and the people becoming monkeys. Then came the Sun '4 Rain', at the end of which everything was burned up and the people became turkeys. Finally came the Sun of our time, called '4 Movement'. In its reign there will be earthquakes and hunger, 'and thus our end shall come', as the *Leyenda de los Soles* says.[11]

In the view of the Aztecs the days all 'work' while they move across the sky. Pictorially the time gods are represented as figures carrying a sum of days, months or years on their backs, following each other in an immense circular sequence. (Interestingly, the ancient Greek poet Hesiod speaks of the hours also as *daimones* – gods.)

9 For details, see Granet.

10 Cf. Saussurè, *Les Origines de l'astrologie chinoise* and *Origine babylonienne de l'astronomie chinoise* 198.5; Needham and Wang Ling, III and elsewhere.

11 Cf. León-Portilla, 37–39.

Our own astrological system consists also of a circular procession of divine images. It stems from Mesopotamia, about the sixth century BC, and also shows some later Egyptian influences. The zodiacal signs originally varied greatly; they were originally earthly, tribal gods which got projected onto the celestial constellations when the Babylonians began to observe the motion of the stars and gather arithmetical information about their movements. Thus the Babylonians began to realize the existence of a *lawful order* in this procession of gods or archetypes over the sky; this order they expressed by numbers. Jung has defined number as *an archetype of order which has become conscious*. The time gods of the Maya and Aztec were also numbers; no wonder that the Chinese too associated the stellar order in the sky with numbers. The word for reckoning, *shih*, is written 示 , the upper horizontal lines representing heaven and the three vertical lines denoting the influence of the sun, moon and stars on the earth. Reckoning was thus closely associated with the prediction of the future.[12] The new Babylonian science-religion spread into Persia and Egypt, where the Babylonian gods were partly renamed and associated with the gods who were known there. The Babylonians also belived in the eternity of the world,[13] and that a certain destiny, Heimarmene, rules in the cosmos.[14] All things on earth correspond to what happens in the skies. In the vast period of 4,320,000 years the whole heavens return to their initial configuration – the Indian myth of the Eternal Return.

These ideas influenced the Greeks from the time of Thales of Miletus onward, and largely stand behind Plato's famous cosmological model of time: above and outside the universe exist the Platonic Ideas, forming one unit around the Idea of the Good. But when the creator god (Demiurge) created the world he could not transfer this model as a whole into the world of transient reality:

> So as that pattern is the Living Being that is forever existent, he sought to make the universe also like it. . . . Now the nature of the Living Being was eternal, and it was impossible to confer this character in full completeness on the generated thing. But he took thought to make, as it were, *a moving likeness of eternity*: and, at the same time that he ordered the Heaven, he made of eternity that abides in unity an everlasting likeness *moving according to number* – that to which we have given the name Time [Aion].[15]

Aion here means an 'aeonic' time[16] which exists in between the timeless world of ideas and the time-bound perishable world of our reality; it consists of long historical aeons. Aion is an everlasting being, the heavenly sphere of the fixed stars, which were thought to be eternal, not subject to suffering and change. It moves in an eternal circle.[17] Only below the moon does the world of Chronos begin, the futile transient 'sublunary' sphere of decay.

The cyclical idea of time is still discussed by physicists, in the form of the so-called ergodic theorem, according to which 'no matter in what state the finite universe may be at a given moment, it will go through all other possible states in a given sequence and come back eventually to the

12 Needham and Wang Ling, 4.

13 Cf. Cumont, 30ff.

14 *Ibid.*, 28

15 *Timaeus*, 37 C, D. See Cornford, 97–100.

16 I use this term following H. Conrad-Martius.

17 Cf. Böhme.

starting state'.[18] Present-day theories about the life of the universe vary widely, however. The most widely accepted is the idea that the universe started with one 'big bang', an explosion, and is expanding towards one end-point of 'heat death'. Other scientists believe rather in a steady-state universe, in which matter is continuously created and destroyed with no beginning or end in sight.

It is certainly not for rational reasons, but on account of the archetypal intuition of a cyclical time (as distinct from the flux aspect) that our invention of clocks made them circular. The faces of clocks today are still mostly fashioned in a circle in order to simulate the heavens.[19]

To the oldest forms of clocks belong the gnomon and sundial, both of which make use of the seeming rotation of the sun around the earth by measuring the shadow of a rod, column or obelisk, and (in the sundial) by tracking its changes of position. The gnomon was probably introduced to the Greeks in the beginning of the sixth century BC.[20] It had the great disadvantage of being able only to indicate local time. Its usefulness was also impaired by the complexity of apparent motion of the sun around the earth. By inclining the rod so as to put it parallel with the axis of the rotation of the earth this was simplified. Now the direction of the shadow was identical for identical hours at a certain place regardless of which day of the year it was; only the length of the shadow continued to vary.[21] This is the sundial which one finds in Egypt from the thirteenth century BC onwards, but probably the Babylonians knew it first. Until the seventeenth century such sundials were more accurate than any mechanical clock.

The invention of the latter goes back to one essential step, to the invention of the toothed gear-wheel in the days of Archimedes, which probably arose from modelling the calendar cycles. Very little is known of it until the ninth century, when in Islamic countries gear-wheels in complicated ratios were in use in astrolabe mechanisms 'recognizable as clockwork ancestors of all modern machinery'.[22]

From such beginnings the mechanical clock slowly evolved. All the widely varying mechanisms that exist contain four basic components, not all of which were discovered at the same time: a motor, first in the form of a weight (later a spring); an oscillating regulator (a pendulum or, as mostly nowadays, an electromagnet),[23] which counterbalances differences of temperature, pressure and shock; an escapement, which compensates for the loss of energy through friction; and a face, indicating the hours, minutes, etc.

But in addition to all the above-mentioned manifestations of a circular idea of time, there also exists from the beginning the idea of a linear and irreversible succession of time, probably based on the observation of the aging of all living beings and of permanent changes brought forth by historical events. Despite their circular models of time, the Chinese, for instance, accumulated a body of historical facts extending over a period of more than three thousand years. They did it in order to study 'how to conduct oneself in the present and the future, how love (yen) and righteousness (i) may result in favourable, evil deeds in unfavourable social results. Thus they believed in a possible moral improvement and social evolution'.[24]

18 Cf. Dauer, 91.

19 Cf. De Solla Price, 'Automata and the Origins of Mechanism and Mechanistic Philosophy', quoted from Haber, 388.

20 Cf. Le Lionnais, 13.

21 Ibid., 14, 19.

22 De Solla Price, 'Clockwork before the Clock and Timekeepers before Timekeeping', 370.

23 Le Lionnais, 26.

24 Needham, Time and Eastern Man, 52–53.

In a similar way the doctrine of the aeons of the five suns of the Aztecs also gives history a linear course, though not leading to evolution but final destruction. Long before man knew about the reasons for aging (because certain cells in our body are not replaced and others are replaced much more slowly), time was in many mythological systems associated with decay and death and even with evil. The Hellenistic Aion was, as we know, also Kronos, the god who eats his children, the former supreme god supplanted by Zeus. This figure of time as devourer even survived into the Christian era, in the symbolic image of Father Time, which unites in itself the attributes of Kronos-Saturn and death. The sixteenth and seventeenth centuries revelled in such macabre representations of the destructive aspect of time.

The Judaeo-Christian tradition believed primarily in a linear model of time, due to the intervention of God and to His Providence, His plan to lead mankind step by step to perfection and finally to the destruction of the world. However, the older Christian notion of time was far from being a purely mathematical parameter. It included certain cyclic elements, as well as the idea of a Divine design – a teleological linearity of time – periodized in terms of the seven days of creation.[25] In the Old Testament there are *typoi* – images or prefigurations of events and things which were later revealed more fully in the New Testament. The Tree of Knowledge, for instance, in Genesis is the same wood from which the rod of Moses was made and also served as a beam in the Temple of Solomon and as the cross of Christ's crucifixion. Thus an eternal pattern interacted with the linear course of history.

Besides stressing that Christ died once and for all, certain Church Fathers, believing in astral influences, accepted in part a cyclical view of history.[26] Such views existed side by side until the seventeenth century. The *illud tempus* model of an extra-temporal existence of all patterns (see p. 12) is also to be found in Christianity in the notion of an *unus mundus*, which was the plan of the cosmos in God's mind before creation. This plan was also called the *Sapientia Dei*, the personified Wisdom of God. Certain primal forms, ideas, prototypes constitute together the *archetypus mundus* or 'exemplar' of the universe in God's mind. It contains a mathematical order which is closely related to the Trinity: number belonging to the Son, measure to the Father and weight to the Holy Spirit.[27] The *unus mundus* is thought of as an infinite sphere, like God Himself.[28] In spite of a repetitive manifestation of the *typoi*, they follow a linear course of evolution in that they make the Godhead and His purpose increasingly manifest: 'What shone forth in the Old Testament, radiates in the New Testament.'

One of the most famous creators of such a pattern of Divine Providence in history is the Abbot Gioacchino da Fiori (twelfth century), who announced that history was divided into three great aeons: the period of the Old Testament, which was the time of the Father, and in which the law and its literal understanding dominated; the first Christian millennium, which was the time of the Son, and in which obedience to the Church and wisdom dominate; and finally the present age of the Holy Ghost in which spiritual men will live in poverty but in complete freedom following the inspiration of the Holy Ghost.[29]

25 Cf. Haber, 385ff.

26 Cf. Eliade, *The Myth of the Eternal Return*, 143–44.

27 Wisdom 11.21. Cf. von Franz, *Number and Time*, 171–73.

28 Cf. Mahnke.

29 Cf. Da Fiore, xli, lii–liii, 23 and elsewhere.

The idea that God had a timeless plan of the world in His mind before its realization in matter is consistent with what has been called the Christian sacramental view of history; it is related to Plato's idea of all things developing out of seeds or primordial archetypes (Ideas). As C. Haber has pointed out,[30] the clock was looked upon in the fifteenth century, as a model of such a divine plan. In 1453, in his *Vision of God*, Nicolas Cusanus writes: 'Let then the concept of the clock represent eternity's self; then motion in the clock representeth succession. Eternity therefore both enfoldeth and unfoldeth succession, since the concept of the clock which is eternity doth alike enfold and unfold things.'[31] Only gradually did this idea of a clockwork universe become desacramentalized – in the eighteenth century the clock became an automaton that had no connection with God.[32]

In physics a breakthrough towards a purely mathematical idea of linear time occurred with Newton, who used a geometrical line to describe measurable time; but its dominant role in physics really came only with the discovery of the Second Law of Thermodynamics, formulated by Carnot and Boltzmann. This theorem says that in every physical process a certain amount of energy becomes irretrievably lost in the form of heat, and that the consequent loss of order – the process known as entropy – will lead to the death of our universe. This led to the idea of the 'arrow of time' in physics, i.e. its irreversible directedness.

However, some physicists believe that mind, in contrast to matter, is a negentropic factor: that it is able, in other words, to recreate order from disorder and build up systems of a higher energy level. This led Olivier Costa de Beauregard even to postulate a cosmic underlying soul or *infra-psychisme*, coexistent with the Einsteinian block universe, as a cosmic source of negentropy.[33] But in modern physics the 'arrow of time' idea still largely predominates.

Besides these developments in physics it was mainly the breakthrough of Charles Darwin's ideas which reinforced the already existing Western predilection for a purely linear model of time.[34] Darwin asserted that all modification of life on earth was mechanical and ultimately due to mere chance. Time thus became a purely mathematical time – a line was sufficient to explain it. Though some 'vitalistic' thinkers have continued to object to this view, it is still a dominant idea in science. Finally, the undeniable subjective psychological changes which man experiences during his life in aging also support the idea of a linearity of time.[35]

Many attempts have been made to bring forth a reconciliation of the linear and cyclic views of time. Thus St Augustine's view of time and eternity is in many ways a combination of both models, just as in another form of the Chinese idea cyclical time was combined with a moral evolution of man through historical experiences.

The image for such a combination is the idea of the spiral or helix. Jung has tried to bring evidence for such a spiralic process within the inner psychic god-image in man, the Self.[36] In the development of this god-man image, as viewed by the Book of Enoch, the Gnostics and certain Western alchemists,[37] the Self was first conceived as a divine Adam figure, and then as a lower earthly Adam figure (after the Fall).

30 Haber, 387ff.

31 Quoted in Haber, 390.

32 Cf. *ibid.*, 392ff.

33 Costa de Beauregard.

34 Cf. Haber, 383, 384, 385.

35 Cf. more to this point, Denbigh, 148.

36 Jung, *Collected Works*, IX, 248–49.

37 Cf. Purce.

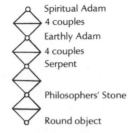

Spiritual Adam
4 couples
Earthly Adam
4 couples
Serpent

Philosophers' Stone

Round object

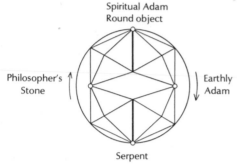

Spiritual Adam
Round object

Philosopher's Stone

Earthly Adam

Serpent

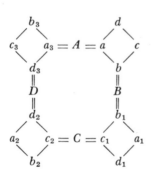

If we wind this chain up in a spiral, we get the picture shown left, above (but one must think that at the upper point the *rotundum* symbolizes a slightly higher level of consciousness than the first Anthropos). This model is, as Jung points out,[38] in line with the main historical development of our idea of God. The still lower counterpart is the serpent which brought about Adam's fall. This corresponds to the 'primary matter' and main concern of the alchemists. When it has been worked through the four elements, it becomes the Philosophers' Stone – another God-Man symbol. The latter was finally conceived as the 'round thing' (*rotundum*), the most basic structure of the universe. In between are four times four centres, which form an upper and a lower marriage-quaternion, the four rivers of Paradise, and the four elements. These quaternions were regarded as revealing the structure of each centre above.

Man was first like a child, dependent on the 'pneumatic' (spiritual) sphere. The latter was threatened by Satan, the dark side of reality, and by man's own instincts (man's own shadow). Christ broke the gates of hell, but did not return as He had promised. The ideas of a quaternion and of the Philosophers' Stone coincide with the beginnings of natural science. Alchemical speculations lead to the idea of four states of aggregation, to the model of a space-time quaternion of four dimensions, and finally to different modern quaternarian models of the subatomic world. The series ends with the *rotundum*, the archetypal image of rotation (which goes beyond the static models of quaternity). This coincides again with the pneumatic Anthropos.

In the form of an equation, this model of the Self could be expressed in the formula shown here:

A stands for the initial state (in this case the Anthropos), *A1* for the end state, and *B C D* for intermediate states. The formations that split off from them are denoted in each case by the small letters *a b c d*. With regard to the construction of the formula, we must bear in mind that we are concerned with the continual process of transformation of one and same substance. This substance, and its respective state of transformation, will always bring forth its like; thus *A* will produce *a* and *B b*; equally, *b* produces *B* and *c C*. It is also assumed that *a* is followed by *b* and that the formula runs from left to right. These assumptions are legitimate in a psychological formula.[39]

What the formula can only hint at is the higher plane that is reached through the process of transformation. ... The change consists in an unfolding of totality into four parts four times, which means nothing less than its becoming conscious.

Jung compares this spiralic process in the Self with the self-rejuvenation of the carbon nucleus in the carbon-nitrogen cycle, where the carbon nucleus captures four protons and emits them again as an alpha-particle at the end of the cycle in order to return to its original structure.[40]

The analogy with physics is not a digression since the symbolical schema itself represents the descent into matter and requires the identity of the

38 Jung, *Collected Works*, 9.II, paras 404, 406ff.

39 *Ibid.*, para. 408.

40 *Ibid.*, para. 410–11. See also Gamov, 72.

41 *Ibid.*, para. 413.

outside with the inside. Psyche cannot be totally different from matter, for how otherwise could it move matter? And matter cannot be alien to psyche, for how else could matter produce psyche? Psyche and matter exist in one and the same world, and each partakes of the other, otherwise any reciprocal action would be impossible. If research could only advance far enough, therefore, we should arrive at an ultimate agreement between physical and psychological concepts. Our present attempts may be bold, but I believe they are on the right lines. Mathematics, for instance, has more than once proved that its purely logical constructions which transcend all experience subsequently coincided with the behaviour of things. This, like the events I call synchronistic, points to a profound harmony between all forms of existence.[41]

This symbolic model of Jung's seems to represent *a basic structure of physical and psychic life*. Interestingly enough, the Melanesian marriage system which attempts to perpetuate the flow of life by a system of marriage exchange between two clans (one matrilinear, one patrilinear) also produces as a result two types of spiral, a closed spiral representing the matrilineal life-line, and an interrupted spiral the virile power.

And last but not least, the genetic substance, as has been discovered by Watson, Crick and Wilkins, is a double helix. Perhaps this model represents a biological analogy to the archetypal idea of time as a spiral, which reconciles the linear and cyclic aspects of time.

Rhythm and periodicity

It was not only Plato who associated time with number. For Aristotle, too, time was a kind of number: 'Time is the number of motion with respect to earlier and later.' But he had only *one* motion in mind; this was the revolving motion of heaven, because it is everlasting and uniform.[1] On the other hand, time for Aristotle does not exist without change:[2] 'Time is a kind of number, in fact the number of continuous movement.' This remark obviously refers to the periodicity of celestial events. This point of view of Aristotle's has been criticized as being only a definition of measurable time but not of time itself. However, it seems to me that the relationship of time and number is in fact a much deeper one. If Alexander Marshak is right,[3] one of the oldest games of palaeolithic man was to make numerical marks on stones and bones as a 'time factoring' occupation. This, according to Marshak, was *the* beginning of our civilization. These marks served from the beginning to keep count of calendrical dates.

The relation of time and number was also a fundamental assertion in China, but with a different nuance. Numbers, in their aspect of quantity, were unimportant to the Chinese; for them number, as Granet points out,[4] was much more a qualitative emblem or symbol, which however, as with us, also denotes regular relations among things. These emblems mirror within a hierarchical order certain basic patterns of the universe; thus they make visible 'the circumstantial individual aspects of the cosmic unity as a whole'. This is where time comes in – because these circumstantial aspects appear in a temporal order. They vary in the course of time, being each a qualitative

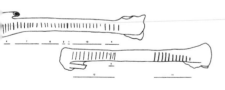

1 Cf. Whitrow, *The Natural Philosophy of Time*, 29, 30 and 'Reflections on the History of the Concept of Time', 4–5.

2 Cf. Ariotti, 75

3 Marshak.

4 Cf. Granet, 154ff., 171ff.

moment.[5] Time in China was therefore a concrete continuum containing qualities or fundamental conditions which can be manifested relatively simultaneously in different places. The famous book of divination, the *I Ching*, is built upon this premise. *Time consists in its view in certain time-ordered phases of transformation of the cosmic whole.*[6] According to the philosopher Wang Fu Ch'ih (1619–92), all existence is a cosmic continuum which is in itself without perceptual manifestation. But on account of its immanent dynamics it differentiates in itself certain images or structures, which follow each other in time, and which can be explained by arithmetical procedures. Thus all numbers in China are also time indicators which tell us something about the quality of each moment. In China, therefore, time never became an abstract parameter or empty frame of reference but was always qualified by the coincidence of the events which all meet at certain of its moments. The whole universe had, in this way, a temporal rhythmic structure. The most basic rhythm is the alternating one of Yang and Yin. Chinese music as a whole was created in accordance with the rhythmical patterns of the universe. An essential idea connected with rhythm was that of enantiodromia: whenever a line of the *I Ching* oracle (see p. 25) or a symbol reached its extreme state of fullness, it jumped over into its opposite. This idea has also been formulated by the Greek philosopher Heraclitus who called fate 'the world order [*logos*] stemming from enantiodromia, the creator of all things'.[7] Jung took the term 'enantiodromia' up again, showing that it was a psychological law; all extreme psychological states tend to tip over into their opposite: goodness into evil, happiness into unhappiness, exaggerated spirituality into surrender to instincts, etc.

The close relation of time with rhythm has contributed to our clockmaking in the form of the pendulum, which was invented by Galileo and perfected for time measurement by Christiaan Huyghens. A simple rhythmic pendulum movement is still used to mark time for music in the form of the metronome. Recently electricity has conquered the field of clockmaking in serving to produce synchronous oscillations.[8] An alternating current keeps up the vibrations and a device regulates the period of the current. Nowadays we have gone even one step further in using piezo-electric quartz or ammonia molecules to produce exceedingly regular and imperturbable oscillations. Compressing a band of quartz produces a migration of some of its electrons towards one of its sides; the process can then be reversed. If the quartz is then connected to an alternating current, the impulses of the electrons are converted into mechanical oscillations of several thousand alternations per second, of such regularity that they even correct the slight inequalities of an alternating electric current.[9] But quartz ages; it is now being supplanted by ammonia molecules of the formula NH_3, the N atoms constantly oscillating towards the opposite pole above a plane of 3H in a frequency of 24,000 megacycles or 24 million vibrations per second. Clocks based on this atomic movement inaugurated the technique of *masers* (*m*icrowave *a*mplification by *s*timulated *e*mission of *r*adiations). *None of these clocks could exist without the basic rhythmicity of energy, i.e., matter.* 'All matter,' says Capra, 'is involved in a continual cosmic dance.'[10] All particles 'sing their song, producing rhythmic patterns of energy.'[11]

5 Cf. von Franz, *Number and Time*, 26.
6 *Ibid.*, 77.
7 Aetius, I, 7.22.
8 Cf. Le Lionnais, 71.
9 *Ibid.*
10 Capra, 241.
11 *Ibid.*, 242.

Modern physics has revealed 'that every subatomic particle not only performs an energy dance but also *is* an energy dance, a pulsating process of creation and destruction'.[12]

Returning to the macrophysical plane and to the bodies of living beings, we still come across the same phenomenon. They all follow certain rhythms, which are now called biological clocks. Plants as well as animals are adapted not only to their spatial environment but also to time: to the solar day through what are called the circadian rhythms, to the cycles of the moon, to the tides of the sea, and even to the solar year. Certain activities, such as looking for food, are not activated by the outer stimulus of sunrise but by an inner rhythm which enables the animal to 'plan ahead'.[13] Plants also 'have something resembling a time memory', for some (not all) begin to open their buds a few hours before sunrise, 'as if they knew that the sun will rise soon'; and if one plunges them artificially into darkness, they still open their buds at the same time of day.[14] The physiological clock in animals seems to work by oscillations.[15] It also works like a master clock from which several other temporarily regulated physiological processes depend. Periods of activity and rest, and quantitative changes in metabolism, temperature and other processes, are regulated in this way. Such rhythms seem to be inherited and are probably endogenous, not produced by outer conditions.[16] In unicellular animals or plants the whole unit is bound to the rhythm. Lowered temperatures (for plants differences of only 5–10°C) slow down the biological clocks.

In more complex beings it is still a matter of discussion how far these rhythms are unified by a regulating organ or spread over different tissues and organs; both seem to be the case.[17] In higher animals there might be a central regulator located in the brain.[18] There also seems to be, as G. Schaltenbrand formulates it,[19] a standardized rhythmic chronological organization in the brain, which functions as a whole.

This basically rhythmic structure of our physiological life is not its only relation to time. As Adolf Portmann has shown, whole patterns of behaviour in plants and animals show a relation to time: 'Every form of life appears to us as a *Gestalt* with a specific development in time as well as space.'[20] The social year of the Samoans and Fijians is calculated according to the cycle of the palolo worm (*Eunice viridis*). Each year this worm sloughs off a part of its body charged with sexual substances and reproduction takes place in the open sea, where the worm reconstitutes itself. This rhythm is related to the phases of the moon; at exactly the same time a certain tree (*Erythrina indica*) blossoms too. Certain sea urchins of the Mediterranean off Egypt, and the oyster and scallop in temperate seas, follow similar rhythms in their reproduction. The migration of birds is also related to the diurnal cycle. The development of animals is

> 'more than a mere undergoing of the temporal process; it is a resistance to [entropic] time, a mode of formation provided in the protoplasm of the particular species. . . . Just as in a well-planned display of fireworks one set piece may bear the next latent within it, so in the life of many insects we find at each stage a prefiguration of new organs, which subsequently unfold in an exactly regulated temporal process.'[21]

12 *Ibid.*, 244.

13 Cf. Bünning, 2.

14 *Ibid.*, 4.

15 Cf. Pöppel, 221.

16 Though also partly working in connection with outer stimuli.

17 Cf. Bünning, 40–41, 52.

18 Cf. Richter, 52.

19 Schaltenbrand, 59.

20 Portmann, 312.

21 *Ibid.*, 313–14.

Man too seems to possess a biological clock or clocks (which gets upset by long-distance flights), although his clock-time consciousness predominates.[22]

In the field of emotion, which is a phenomenon very much on the border between physiological and psychological reality, rhythm is essential. In states of strong emotion we make rhythmical movements (stamping our feet, for instance) and tend to repeat endlessly the same thoughts and utterances. This led Jung to suspect that unconscious complexes might have a periodical rhythmic nature, a fact which is now under observation through the study of the periodic recurrences of themes in dream series.[23] The great common complexes of mankind, which Jung called the archetypes, manifest as physical and psychic patterns of behaviour and were in the past experienced as gods or mythic images. The astrological systems discussed above are in fact attempts on mankind's part to express a temporal procession and recurrent order or 'play of the archetypes', constituting an aeonic time rhythm. That such a play exists may well be true, but we are far from understanding it. The relation of gods to moments and to chance, which I discuss in the following sections, also points to a partly timeless and partly timebound nature of the archetypes.

What has been much more studied lately by psychologists is the immense variety of man's subjective time. Thus, for instance, young people live more in expectation of the future, old people look more at the past. Aging people generally feel that time goes faster, probably due to their own slowing down.[24] 'Subjective' time differences might be based largely on physiological factors (for instance temperature), which have been shown to interfere with the speed of our intake of sensory information and our estimate of time. Intoxication with hashish, opium, mescalin, etc., also expands or condenses our subjective experience of time.[25]

Subjective time does not only vary from one individual to another, but also between different social, ethnic and psychological groups. Adaptation to clock-time remains (in my experience) difficult for intuitive people, while people whose sensation function is dominant get so stuck in what is now that they are often incapable of imagining a change tomorrow. Because time is connected with our whole 'rhythm of life', the adaptation to it is generally disturbed in neuroses (and even more in psychoses). Some people, for instance, live 'faster' than they can afford, or they lag behind in their own inner development. Then they feel 'haunted by time'.

If it is true, as I have tried to show, that time is closely related to the rhythm of the inner god-image, the Self (i.e. the conscious-unconscious totality of the psyche), then it is obvious that every neurotic deviation from the rhythm of the Self entails also a disturbed relation to time.

Besides the biological rhythms, memory processes play a role in our subjective time experience. Their complex nature has not yet been elucidated. Many scientists think that 'the brain or the mind retains a complete record of the stream of consciousness, that is of all details which were recorded mentally (including infraconscious awareness) at the time of occurrence, although later most of it is entirely lost to voluntary recall'.[26] Many things we can recall 'are only generalizations and summaries', that is, most memories seem to become part of a more or less extensive

22 Cf. Richter, 39, 52, and the litera-ture cited there.

23 By Paul Walder, C.G. Jung Institute, Zurich. Cf. also, on the periodicity in the behaviour of schiz-ophrenics, Aaronson, 308.

24 Cf. Kastenbaum, 20ff, esp. Green, 1ff.

25 Cf. Le Lionnais, 103, 107.

26 Cf. Whitrow, *The Natural Philosophy of Time*, 105.

27 *Ibid.*, 107.

28 Quoted *ibid.*, 111.

29 *Ibid.*, 11.

organization, called by Bartlett 'schemata',[27] and which Suzanne Langer called 'man's symbolic transformation of experience'.[28] From a Jungian point of view this organization would be produced by the archetypes, which are inborn psycho-physical ordering principles of human experience. Whatever role memory plays in it, our mind, says Whitrow, 'is certainly temporal in its very nature. It manifests in our consciousness as a "train of thoughts"'.[29] Here we return to the primitive notion of time as a quantitative and qualitative stream of simultaneous outer and inner events. In his *Experiment with Time*, J. W. Dunne has tried to formulate a model of multidimensional time, mirroring psychic states and their coincidental 'times'. This model has not been generally accepted; but it seems to me that Dunne is basically correct, in seeing time as a multidimensional phenomenon which is characterized by the simultaneity of different psychological conditions.

Necessity, chance and synchronicity

Causality has been accepted in some form in all civilizations. In the Far East its basic form is the concept of *karma*. Throughout all the innumerable reincarnations of a person, a certain identity goes on, in a chain of causation carried by *karma*: 'If we see an identity of being in events in the course of time, it is due to the causal chain that connects them.'[1]

What we call causality in the West has its roots in the Greek images of Ananke (Necessity), Dike (Justice), Heimarmene (Allotted Fate), and Nemesis (Retribution) – all goddesses which were feared and respected. They were responsible for the balanced play of opposites in the universe: 'The source of generation for all things is that into which their destruction also leads . . . according to Necessity, for they pay penalty and retribution to each other for their injustices according to the order of Time,' says Anaximander.[2] And Heraclitus stressed that 'all things happen by strife and necessity'.[3] Later, in the philosophy of the Stoics, Ananke or Heimarmene became *the* all-ruling world principle, which even rules over the gods. According to the Orphics, Chronos (Time) was mated to Ananke (Necessity), which holds the universe in powerful fetters, surrounding it in the form of a serpent.[4] Our word 'Necessity' itself is related to the Latin *necto* ('I bind'), *nexus* ('bound'). This inexorable goddess personified also the fetters of death, our destiny (*destino* also means 'I bind'). Ananke spins the thread of our life and cuts it at its fated end.

In the Christian era the concept of Necessity did not disappear, but was projected on to the lawful order of nature, created by God Himself (Who also, however, sometimes interfered with it through miracles). Only with René Descartes (1596–1650) did the principle of determinism, in the form of general natural laws, become absolute, excluding any possible new creative divine interference: 'And generally we can assert that God does all that we can understand, but not that He cannot do what we cannot understand.' He *could* act differently but He does not want to do so. *God's activity coincides completely with the principle of causality.*[5] Much the same is true for Isaac Newton. According to him God created in the beginning the material particles, the forces between them and the fundamental laws of motion, and it has continued to run ever since like a machine governed by

1 Quoted from Watanabe, 267ff.

2 Anaximander, 'Simplicius In Aristotelis physicorum libros Commentarius'. Quoted in Diels, 476.

3 Frgm. 80, from Clement of Alexandria. Cf. for this Ariotti, 71.

4 Cf. Onians, 251, 332.

5 Cf. Barth; von Franz, 'The Dream of Descartes', 84f.

immutable laws.[6] It was an easy next step to exclude the idea of God altogether and thus, in the age of materialism, the universe became an immense mechanical clock which ticks on stupidly into all eternity.

From that time on, the belief in the absolute validity of causality lasted until the beginnings of quantum physics, where the study of elementary particles forced the physicists to replace it with the concept of mathematical probability. Certain prediction is no longer possible for the behaviour of single particles, but only for very large sets of particles. However, this does not simply reflect our ignorance of the physical situation, as does the use of probabilities by insurance companies. As Capra formulates it,[7]

> 'In quantum theory we have come to recognize probability as a fundamental feature of the atomic reality. . . . Subatomic particles do not exist with certainty at objective places, but rather show "tendencies to exist", and atomic events do not occur with certainty at definite times and in definite ways, but rather show "tendencies to occur",'

There thus exists a certain margin of uncertainty.[8] Einstein could not at first accept this and uttered to Niels Bohr his famous words: 'God does not play with dice!'

Quantum physics came across another fact which concerns the problem of time even more directly, namely, the so-called symmetry concerning the direction of time. In the space-time diagram here, this can be read either as an electron-photon collision or scattering (the electron being depicted by an upward arrow, the photon by a broken line) or as a positron-photon scattering (the positron being depicted by a downward arrow). 'The mathematical formalism of field theory suggests that these lines can be interpreted in two ways, either as positrons moving forward in time or as electrons moving backwards in time.'[9] This feature of the world of subatomic particles can also be sketched like the third diagram: an electron (solid line) and a photon (broken line) approach each other. At point A the photon creates an electron-positron pair, the electron moving on to the right, the positron to the left. At point B the positron collides with the initial electron and they annihilate each other, creating a photon which flies off to the left. 'We may', however, 'also interpret the process as the interaction of the two photons with a single electron travelling first forwards in time, then backwards and then forwards again.'[10] We can therefore interpret the process as a four-dimensional pattern of interrelated events which does not have any definite direction of time attached to it.[11]

In spite of this, the 'arrow of time' and causality are still valid in many areas of the world of matter. In order to find a more general framework for the description of the protons and neutrons (the most basic form of particles) one approach is to use a so-called S-matrix, first proposed by Werner Heisenberg.[12] The circle represents simply the area where complicated single observed processes might take place. A and B are two particles (of any kind) which undergo in this circle a collision-process and emerge as C and D, two different particles.[13] This S-matrix theory bypasses the problem of specifying precisely the position of individual particles.

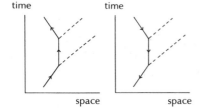

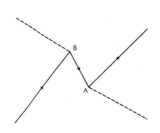

6 Adapted from Capra, 56.

7 *Ibid.*, 68.

8 Capra, 68.

9 *Ibid.*, 182–83.

10 Capra, 184.

11 *Ibid.*, 185.

12 *Ibid.*, 262ff.

13 Cf. Capra, 262–63.

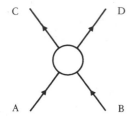

C   D

A   B

The use of the S-matrix involves a number of basic principles.[14] The first is that the reaction probabilities must be independent of displacements of the apparatus in space and time, and independent of the state of motion of the observer. The second is that the outcome of a particular reaction can only be predicted in terms of probabilities. The third principle is related to causality: it states that energy and momentum are 'transferred over spatial distances only by particles and this occurs in such a way that a particle can be created in one reaction and destroyed in another only if the latter reaction occurs *after* the former'.[15] There is a fourth factor (Capra includes it in the third): it concerns the values at which the creation of new particles becomes possible (although not predictable). At those values the mathematical structure of the S-matrix changes abruptly: 'it encounters what mathematicians call a "singularity"'. The fact that the S-matrix exhibits singularities is a consequence of the causality principle, but the location of the singularities is not determined by it.' (See p. 28.)

The logical opposite of the concept of causality (and its historically older form, Necessity) is *chance*. The latter seems to reach back to older religious ideas and habits of life than the former. As Hermann Usener has shown,[16] the Romans and Greeks possessed in their pantheon many time-gods, in the sense that certain gods personified specific instants in time. There was a god of the moment when the horses panic, of the right moment to pull up weeds, to take the honey out of the hive, etc. Hermes was, among other things, the god of that moment when a sudden silence fell upon a social gathering. One god, Kairos, who was iconographically related to Hermes, was especially important: he personified a lucky coincidence of circumstances, favourable for action; one had to 'grasp Kairos' (one's chance) by the hair, otherwise he escaped. Another 'lucky-chance' deity was Nike (Victory). She represented that mysterious agent or moment when the scales tip in favour of one or the other combatant in war or competitive games. Nike was the daughter of Styx, the circular river in the Underworld, closely related to Oceanos, the time-river god. Still another time-goddess was Fortuna; she was depicted with another time symbol, the wheel.

The great pre-scientific attempts to explore the numinous quality of a moment were the astrological systems, mentioned earlier. Every day, month, year and aeon had its own 'god' or divine symbol, lending man's essence, action and life a definite quality. The astrological time-gods, however, were no longer pure chance-gods, for they moved *in an ordered temporal succession* over the skies in the course of that time-ordered play of archetypes, the laws of which we do not know.

These gods do not really belong to the sky: they have been projected on to it by man. However, they *do* belong to time. This is made evident by the fact that Chinese, Aztec, Maya and Western astrological doctrines developed a technique *by which one could make the same predictions as does astrology by an earthly numerical oracle*. Today one of them, the *I Ching*, has become famous. According to it one can determine the meaning of a given moment by counting off by fours a randomly picked bundle of 49 yarrow stalks. The remainders constitute four types of line, two masculine (—) and two feminine (– –). Six such lines (two trigrams or *Kua*) constitute an oracle answer. There are 64 double-trigrams, depicting basic symbolic life situations within

14 *Ibid.*, 274–76.
15 *Ibid.*, 275.
16 Usener, 279ff.

the moving on of Tao. In the West there exists a similar technique called geomancy, whose symbolic configurations of four lines, each of either two or one dot, is constituted by counting off in pairs a randomly picked cluster of pebbles or dots. While the *I Ching* oracles have been elaborated into a deep philosophy of existence, Western geomancy has largely remained a primitive divination technique. But not so in West Africa, where it is also connected with a differentiated religious system.[17] Its efficiency is based on the workings of a divinity called Fa, who has no collective cult but only talks 'individually to the individual'. He is the god of truth, but he uncovers his whole secret to us only in the after-life. He is not a force of nature but symbolizes God's solicitude for his creation. He is the 'Lord of Life' and 'the hole that calls us over to the Beyond'. Through the geomantic oracle he communicates to the medicine-man who handles it the inner truth of every situation.

Looked at from a mathematical point of view these divination techniques form a complementary opposite to the calculus of probability. The latter becomes more accurate as more cases are considered; its answer is never yes or no but a fraction between 0 and 1 (no and yes). Chance is implied but eliminated as much as possible. On the other hand, the divinatory oracles of the *I Ching* and of geomancy operate with whole (natural) numbers. Their answer is based on a just-so absolute numerical result. Chance is taken *into the centre of attention*; repetition and averages are set aside.[18]

By studying the *I Ching* over many years, Jung was inspired by it to seek for a new principle complementing our Western concept of causality, a new principle which he called synchronicity. In his Introduction to the *I Ching* he writes:

'The Chinese mind, as I see it at work in the *I Ching*, seems to be exclusively preoccupied with the chance aspect of events. What we call coincidence seems to be the chief concern of this peculiar mind, and what we worship as causality passes almost unnoticed. We must admit that there is something to be said for the immense importance of chance. An incalculable amount of human effort is directed to combating and restricting the nuisance or danger represented by chance. Theoretical considerations of cause and effect often look pale and dusty in comparison to the practical results of chance. It is all very well to say that the crystal of quartz is a hexagonal prism. The statement is quite true in so far as an ideal crystal is envisaged. But in nature one finds no two crystals exactly alike, although all are unmistakably hexagonal. The actual form, however, seems to appeal more to the Chinese sage than the ideal one. The jumble of natural laws constituting empirical reality holds more significance for him than a causal explanation of events. . . . The manner in which the *I Ching* tends to look upon reality seems to disfavour our causalistic procedures. The moment under actual observation appears to the ancient Chinese view more of a chance hit than a clearly defined result of causal chain processes. *The matter of interest seems to be the configuration formed by chance events in the moment of observation* . . . synchronicity takes the coincidence of events in space and time as meaning something more than mere chance,

17 Cf. von Franz, *Number and Time*, 265ff. From Maupoil.

18 Von Franz, *Number and Time*, 220, 223.

namely, a peculiar interdependence of objective events among themselves as well as with the subjective (psychic) states of the observer or observers.

'The ancient Chinese mind contemplates the cosmos in a way comparable to that of the modern physicist, who cannot deny that his model of the world is a decidedly psycho-physical structure. The microphysical event includes the observer just as much as the reality underlying the *I Ching* comprises subjective, i.e., psychic conditions in the totality of the momentary situation. Just as causality describes the sequence of events, so synchronicity to the Chinese mind deals with the coincidence of events. The causal point of view tells us a dramatic story about how *D* came into existence; it took its origin from *C*, which existed before *D*, and *C* in its turn had a father, *B*, etc. The synchronistic view on the other hand tries to produce an equally meaningful picture of coincidence. How does it happen that *A'*, *B'*, *C'*, *D'*, etc., appear all in the same moment and in the same place? It happens in the first place because the physical events *A'* and *B'* are of the same quality as the psychic events *C'* and *D'*, and further because all are the exponents of one and the same momentary situation. The situation is assumed to represent a legible or understandable picture.'[19]

Besides experimenting with the *I Ching* Jung observed that frequently a patient would dream of symbolic images which then in a strange way coincided with outer events. If one looked at the latter as if they were symbols, they had *the same meaning* as the dream images.[20] This seems mostly to happen when an archetype is activated in the observer's unconscious, producing a state of high emotional tension. In such moments psyche and matter seem no longer to be separate entities but arrange themselves into an identical, meaningful symbolic situation.[21] It looks at such times as if *physical and psychic worlds are two facets of the same reality*.

This unitary reality Jung called the *unus mundus* (the *one* world).[22] Synchronistic events are, according to Jung, sporadically and irregularly occurring parapsychological phenomena. But they seem to be only special incidences of a more general principle which Jung termed *acausal orderedness*.[23] The latter means that certain factors in nature are ordered without its being possible to find a cause for such an order. Within the realm of matter this would be such facts as the time rate of radioactive decay, or the fact that the speed of light is 186,000 miles per second and not more or less.[24] In the realm of the mind or psyche acausal orderedness is manifest in such examples as the fact that 6 is a perfect number; the addition of its factors, $1+2+3$, and their multiplication, $1 \times 2 \times 3$, both yield 6. We are mentally forced to accept this as true without being able to indicate a cause for 6 having just this quality. These orders can be studied and underlie the above-mentioned divination techniques. As opposed to them, synchronistic events form only momentary special instances in which the observer stands in a position to recognize the third, connecting element, namely *the similarity of meaning* in the inner and outer events. Their orderedness 'differs from that of the properties of natural numbers or the discontinuities of physics in that the latter have existed from eternity and occur regularly,

19 Jung, *Collected Works*, I, iiiff (my italics).

20 Jung, *Collected Works*, VIII, 870, 902ff, 915; von Franz, *Number and Time*, 6ff.

21 Cf. also von Franz, *Number and Time*, 7.

22 *Ibid.*, 9.

23 Jung, *Collected Works*, VIII, 964–65.

24 Naturally one could use another measure, but the just-so-ness would remain.

whereas synchronistic events are *acts of creation in time*.'[25] In this sense the world of chance 'would be taken partly as a universal factor existing from all eternity and partly as the sum of countless individual acts of creation occurring in time'.[26]

These creative acts in time, however, do not occur completely outside recognizable means of prediction, but on the contrary take place within certain fields of probability within the acausal orderedness.[27] It is these fields of psycho-physical probabilities which the divination techniques try to explore by means of *numerical* procedures.

Since Jung published his discovery, some recent developments in nuclear physics seem to me to have come closer to similar ideas. In the S-matrix theory, mentioned above, it is held that within the causal and predictable chain of events observed in the channel of sensation there happens also the *unpredictable* creation of the new particles which are called 'singularities'. There is, however, a difference between the observation of meaningful synchronistic events and the physical 'singularities' for which we cannot detect any *psychological* meaning. Jung has proposed referring to the synchronistic events in which no observer can state the meaning as *similarity*.[28] This coincides with L. L. Whyte's investigations,[29] according to which in nature 'incomplete patterns are trying to become complete'. The mathematical symbolism of patterns displays a tendency and movement of its own: towards completion. This is true not only for crystalline forms but also for microphysical patterns. All incomplete structures are in some degree unstable and tend either to complete themselves or to disintegrate.[30] Thus Whyte defines life as a spreading of a pattern as it pulsates.[31] One could interpret this 'spreading' as being based on 'similarity'. But only the human mind can see meaning in this and can consciously experience the oneness of mind and matter.

Modern physicists have come independently to a similar idea of a basic oneness of the cosmos (realizing simultaneously that all the things we can say about it are constructs of our own mind). In this 'one world', as Fritjof Capra puts it, 'every particle consists of all other particles',[32] and at the same time they also all 'self-interact' by emitting and reabsorbing virtual particles.[33] 'Particles are not isolated grains of matter but are probability patterns, interconnections in an inseparable cosmic web.'[34] They are various parts of a unified whole.[35]

What is different in this physicist's 'one world' from Jung's *unus mundus* is that the latter also includes psychic reality, or rather that it *transcends both psyche (mind) and matter*. The ultimate nature of both, the *unus mundus* itself, is transcendental; it cannot be grasped directly by our consciousness.[36] Synchronistic events are 'singularities' in which the oneness of psyche and matter, the *unus mundus*, becomes sporadically manifest. Number too seems to be in an exact relation to both realms, being an aspect of all energy manifestations and of the reasoning working of our mind. Number, according to Jung, is the most basic or primitive form of the archetypes, which are the 'arrangers' of our conscious ratiocinations;[37] 'it is quantity as well as meaning'.

But where is time in all this? Let us return briefly to the two Chinese time-mandalas, the Sequences of Earlier Heaven and Later Heaven. The oldest

25 Jung, *Collected Works*, VIII, paras 964ff (my emphasis), also note to para. 518.

26 *ibid.*, paras 964ff.

27 Von Franz, *Number and Time*, 12.

28 Jung, *Collected Works*, VIII.

29 Whyte, 97–98.

30 *Ibid.*, 101.

31 *Ibid.*, 104ff.

32 Capra, 295.

33 *Ibid.*, 244.

34 Quoted *ibid.*, 203.

35 *Ibid.*, 159; cf. also 137–38.

36 Cf. Jung, *Collected Works*, VIII, paras 420, 439.

37 Cf. von Franz, *Number and Time*, 9, 15.

38 *Ibid.*, 241ff.

39 Granet, 316, 321.

40 Eigen and Winkler.

Chinese shamans drew their equivalents, the Ho-t'u and the Lo-shu number patterns on two boards, round and square respectively, ran a stick through the centre of both and spun them around it. Where they stopped, one above the other, he 'read' out the symbolic situation in time. The interplay of the two boards was understood as a sacred marriage between Heaven and Earth, the coming together of the eternal order of time with the actual just-so moment, indicating 'fields of probability' within which synchronistic events could occur.[38] The Earlier Heaven corresponds to what Jung called acausal orderedness; it is timeless. The Later Heaven deals with the lapse of time. Time in it is a 'field' which imparts to all things which coincide within it a definite quality. It mediates, as the Hopi see so clearly, between the possible and the actual meaningful chance-event. Between the two realms stands man, who sets the boards in motion. Here chance or freedom enters the game and the laws which govern it, though man is naturally also part of the whole coinciding situation. The Chinese saw the wise man's relationship to the cosmos as a ritual play. His

'superiority and freedom is founded in rites which are wholeheartedly played by the player. ... It is expected that a deeply serious and straightforward game will mediate clarity or wisdom and bring about liberation. The rites call for sincerity; the game requires fixed rules, or at least the prototypes of rules.'[39]

This means that the rules are not absolute laws, for they preserve room for play.

Interestingly, the Nobel Prize winner Manfred Eigen has recently also made an attempt to explain evolution and biological processes by comparing them with number games.[40] However, he does not yet believe in a meaning in chance, only in 'blind' chance. But this notion of 'blind' chance is a remnant of the age of a deterministic view of reality; it is perhaps only 'blind' when *we* are blind to its meaning.

Transcending time

We have seen that the image or notion of time nearly always contains several pairs of opposites, or even triads. In China we have a timeless order, a cyclic time-order, and a linear historical time. In India Brahma is time and not-time. The Maya draw time, *kin*, as an image which contains a static element, the flower, and a flowing element, the arrows of the sun. The Aztecs know of a cyclical time and a linear historical time of five sun-periods. Plato's system contains a timeless world of Ideas, a cyclic aeonic time and a perishable world of ordinary time. The old Persians had two Zurvān figures: Infinite Time and Time of Long Dominion (aeonic time). In his excellent survey of the different philosophical notions of time, G. J. Whitrow states that basically some always tend to eliminate time, others to take it as a basic objectively existing factor. Time is seen as life and death, good and evil. Basing himself on these facts (and others) J. T. Fraser attempted to define time in terms of conflict. I would prefer to apply Nicolas Cusanus' definition of God also to time, that it is a coincidence of opposites: *coincidentia oppositorum*.

Let us briefly turn to the extremest of all complementary opposites: to the contrast of time and not-time. The greatest efforts to transcend time have been made by the Eastern sages, for instance in the practices of Indian yoga. However, when a yogi seeks to transcend time he does not do so with a jump. Through his breathing exercises he tries initially only to overcome ordinary time and 'burns up' all his karmic personal inheritances. Then he begins to breathe according to the rhythm of the great cosmic time.[1] His inspiration corresponds to the course of the sun, his expiration to that of the moon: 'The yogi lives a cosmic Time, but he nevertheless continues to live in time.' Later he attempts to unify even these two rhythms and thus abolishes the cosmos and unites all opposites. He breaks the shell of the microcosm and transcends the contingent world, which exists in time. The ultimate foundation of reality, into which he breaks through, is both time and eternity; what we have in fact to overcome is only our wrong assumption that there is nothing outside ordinary time.

In Taoist mysticism, and in Zen Buddhism, we find very similar ideas. The *Lankavatara Sutra*, for instance, says:

'Why are all things neither departing nor coming? Because though they are characterized with the masks of individuality and generality, these masks coming and departing neither come nor depart. . . . Why are all things permanent? Because though they take forms . . . they take really no such forms and in reality there is nothing born, nothing passing away.'[2]

'Being and non-being – between these two limits the mind moves; with the disappearance of this field the mind properly ceases to operate. When an objective world is no more grasped there is neither disappearance or non-being, except something absolute known as Suchness (*tàthatāvastū*) which realm belongs to the wise.'[3]

The ancient *Treatises of Seng-Chao* elucidate this Buddhist idea in even more detail concerning time:[4]

'When [the Sutras] say that [things] pass, they say so with a mental reservation. For they wish to contradict people's belief in permanence. When they say that things are lost, they say so with a mental reservation, in order to express disapproval of what people understand by "passing". . . . Their wording may be contradictory, but not their aim. It follows that with the sages: "Permanence" has not the meaning of staying behind [while the Wheel of Time, or Karma, moves on]; "Impermanence" has not the meaning of outpassing [the Wheel]. . . . People who seek in vain ancient events in our time conclude that things are impermanent; I who seek in vain present events in ancient times know that things are permanent. . . . [Therefore] the Buddha is like the Void, neither going nor coming. He appears at the proper moment but has no fixed place [among beings]'.[5]

How one can exist both in ordinary time and in aeonic time together can best be illustrated by the story of the death of the great Zen Master Ma. When he reached the end of his life and was lying very sick in his room, the warden of the monastery visited him and asked reverently: 'How has the

1 For this cf. Eliade, 'Time and Eternity in Indian Thought', 197–98.

2 Suzuki, 304.

3 *Ibid.*, 306.

4 Chao-Lun. Seng-Chao belongs to the period of Buddhism before the T'ang period, 49.

5 *Ibid.*, 110.

Venerable One's state of health been recently?' Ma replied: 'Buddha with the sun visage, Budddha with the moon visage.' As Wilhelm Gundert explains, these words hint at a passage in the Third Sutra of the Name of Buddha, where it is explained that the life-span of the Buddha with the moon visage is only one day and one night. The life-span of the Buddha with the sun visage is one thousand eight hundred years. Both Buddhas, however, are only facets of the Great One.[6] After one day and one night Ma died. His mortal part (his moon visage) lasted only that long, but another, more archetypal, part of himself was to last much longer; and beyond it there would even be an eternal kernel; but of this Ma did not speak, because it is ineffable.

In a certain way – and this has been visible in earlier sections of this book – one could clarify the problem of time and not-time in the following way:

— Ego-time

— aeonic time

— illud tempus

— timeless centre

One could compare time to a rotating wheel: our ordinary communal time, which we are aware of in our ego-consciousness, would be the outermost ring which moves more quickly than the others. The next inner ring would represent aeonic time, moving progressively more slowly as the centre is approached. This aeonic time is represented in the idea of the Platonic Year or the Aztec ages or Suns – a time which lasts infinitely longer than our ordinary time. The next and smallest would represent Eliade's *illud tempus*, which is right on the razor's edge between time and no-time, representing, as he says, an 'extratemporal moment of creation'. It is right between unutterable eternity and the beginnings of aeonic time, the latter being the slow-moving life of the archetypes. And finally there is the hole, the non-rotating empty centre of the wheel, which remains permanently quiet, outside movement and time. This is, for instance, the Chinese Tao which lies beyond the rhythms of Yang and Yin:

> There was something formlessly fashioned
> That existed before Heaven and Earth,
> Without sound, without substance,
> Dependent on nothing, unchanging,
> All-pervading, unfailing.[7]

It is completely void and still:

> Push far towards the Void,
> Hold fast enough to Quietness.
> . . .
> This return to the root is called Quietness,
> Tao is forever and who possesses it,
> Though his body ceases, is not destroyed.[8]

6 Bi-Yän-Lu, 97.
7 Waley, ch. 25.
8 *Ibid.*, ch. 16.

Western mysticism also knows of this step of completely transcending time at the moment of union with the extratemporal godhead. One of the great masters who stressed this point often is Meister Eckhart. He says:

'St Paul says: "In the fullness of time God sent his Son." . . . It is the fullness (or end) of the day when the day is done. . . . Certain it is that there is no time when this birth befalls, for nothing hinders this birth so much as time and creature. It is an obvious fact that time affects neither God nor the soul. Did time touch the soul she would not be the soul. If God were affected by time he would not be God. Further, if time could touch the soul, then God could not be born in her. The soul wherein God is born must have escaped from time, and time must have dropped away from her.'[9]

'Suppose that someone had the knowledge and the power to sum up in the present now, all the time and the happenings in that time of six thousand years, including everything that comes until the end, that would be the fullness of time. That is the now of eternity, where the soul in God knows all things new and fresh.'[10]

But from the timeless God flows the 'flow of grace' which creates an ever-present now – so that God is simultaneously stillness and an everlasting flux.

9 Pfeiffer, I, 80–81. Cf. also 227.

10 *Ibid.*, II, 152; also 144, 155, 186, and I, 399.

1　In classical China the Dragon symbolizes the creative dynamic force in the universe, the male Yang principle which acts in the world of the invisible, with spirit and time as its field (while the female Yin principle acts upon matter and space). Yang creates the beginning; Yin, the completion. In the centre can be seen the primordial Pearl of Beginning, from which all things come forth. The meaning of time is that, in it, stages of growth can unfold in clear sequence. Heaven (or Yang) shows a powerful and ceaseless movement that by its nature causes everything to happen in one single, synchronous time; and this, from the divine standpoint, is destiny. (Dragon badge from an Emperor's surcoat, China, 1850–75.)

2 Oceanos flows as a great river round the rim of the earth, the psyche of the universe and 'the generation of all'. He is Time (Chronos) itself, and also Aion: the power controlling the changes in the world, the mystical 'round element' (or Ω omega) which in late antiquity symbolized the 'lifetime', a 'period of time', and 'eternity'. He was also represented as a snake eating its tail, surrounding the firmament as the path of the sun, carrying the signs of the Zodiac on its back. As the boundary of the world, Oceanos is its compelling destiny. (Roman relief, the so-called Bocca della Verità, Santa Maria in Cosmedin, Rome.)

3 Time, being experienced as a constant flow of inner and outer events, was conceived in antiquity as a river. All river gods were pictured as male figures, often with the horns of a bull, because they symbolized dynamic forces. Eridanos was the son of Oceanos-Chronos (Time) and the sea-goddess Tethys. According to Virgil, he wells up from Elysium, the timeless land of bliss. In the dreams of modern people, time is still often represented by a river, and many devices to measure time, such as the ancient water clock and the hourglass, use the flow of a substance to do so. (Eridanos, from a Latin astronomical manuscript, c. AD 1000.)

ORNAMENTO · CELI:

4 According to Christian tradition, God – before creating the actual universe – conceived a mental model of it, in which all the arche-types or 'germs' of things to come existed simultaneously. This model was a timeless *unus mundus* (one world). In the picture above, God rests on this spherical model. Only after He created the actual world (on the right) and in it the sun and moon – i.e. day and night – did time come into existence. In the Judaeo-Christian tradition time is thus not cyclical but reveals its meaning in the irreversible course of God's relation to man and man's salvation. (The Creation, mosaic, Duomo, Monreale, Sicily, before 1183.)

5   In the centre is the Pearl of Beginning, the germ of the universe. In this Primal Monad appears the *t'ai chi*, the great ridge-pole, the oneness which creates the duality of the two basic rhythm of the universe: Yang (light, heat, hardness, expansion, masculinity) and Yin (shade, cold, softness, contraction, feminin-ity). These two points set fixed limits of change: 'Reversal is the movement of Tao.' The timebound inter-action of these rhythms gives rise to all phenomena in the universe. The bird, the phoenix, belongs to K'un the mother, Earth the Receptive; the clouds belong to Ch'ien the dragon, Heaven the Creative. In the four corners are four of the eight Kua or world principles: above (left and right) Ch'ien and K'un; below (left and right) Dui, the lake, the Joyous, and Sun, the wind, the Gentle. (Embroidered robe of Ch'ing Dynasty, China, 18th c.)

8   In ancient Egypt the sun god Ra
was the lord of time, because he
sets its measures when he sails in his
barge over the upper and lower
firmament. Each hour he assumes
the shape of a different animal god.
Time thus acquires changing qualities and is made visible in an endless
circular 'procession of archetypes'.
(The boat of Ra, scene from the
Book of Gates, papyrus, Egypt.)

9 In the West, time is represented as a snake, and in the East as a dragon; in the New World it is symbolized by a snake with two heads, one representing life and the other death. Whilst in China the play of the opposites is viewed as being harmonious, the idea of the Mexican double serpent is more tragic (the heads look in opposite directions). Time brings joy, luck and life, but also misfortune and death, and the whole temporal world will come to an inevitable catastrophic end. (Double headed serpent in turquoise mosaic, Mixtec workmanship, Mexico, 13th–14th c.)

CVALERI
VSHERACLESPAT
ETCVALERI I
VITALISFENICO
MESSACERDO
TES SPCPSR
D DIDIACVIMP
COM
VVET
EPTI
MIANO
COS

10   In the Mithraic mysteries of late antiquity, Aion, holding a key and/or a sceptre, is the keeper of the gate. He represents time, but also long periods of time and eternity. His lion head denotes the summer and his fiery nature; the snake his wintry and moist aspect. Often his body or the snake are marked with the signs of the zodiac. Prayers invoke him as the world-soul, as an all-embracing spirit, light and dark, ruler of everything. To the initiate, he is the Lord of Light who opens the gates of the Beyond. (Aion, Mithraic cult figure, late Roman.)

11   The ancient Greeks, Romans and Slavs worshipped certain gods who personified different numinous 'magical moments' in time. One of them is Nike (Victory), daughter of Styx, the river of the Underworld. She symbolizes that mysterious moment when a game or battle turns into victory. She is always represented winged and in swift movement, because she appears and disappears so suddenly. The image of Nike giving a hero a wreath of victory was copied on early Christian sarcophagi and became an image of an angel who crowns the dead with the crown of victory over death. (Nike, acroterion from early stoa of Zeus, Athens, c. 400 BC.)

12 Tibetan picture of the Wheel of Sangsara, of Becoming. The three animals in the centre, pig, cock and snake, symbolize passion (desire), hatred, and unconsciousness (stupidity). They keep the Wheel moving. The figures to the right are descending into Hell and are being tortured by demons. On the left people are ascending; on top is a naked yogi, carrying the banner of victory. He is about to escape from the Wheel and leave the world of Karmic existence for ever. The six segments depict six destinies visited by Avalokitesvara on his mission of salvation. The monster embracing the Wheel is Anityata (Impermanence), which devours all existence. (Wheel of Life, *t'angka* painting, Tibet, 18th–19th c.)

13 Another goddess of a magical moment of time is Fortuna, the goddess of good and bad luck. Originally she was one of the pre-Roman great mother goddesses (*matronae*) and had the wheel as her symbol. This wheel was later interpreted as the wheel of time, which carries some people upward towards luck and success and throws others down into misery. Sometimes Fortuna is depicted as blindfold, because her choice of lucky and unlucky people is not concerned with ethical conduct. (Wheel of Fortune, from ms. of John Lydgate's *Troy Book and Story of Thebes*, England, c. 1455–1462.)

14   Before they became the heavenly images of constellations, the figures of the Zodiac were earthly tribal gods. In Mesopotamia they were projected on to the sky, and as early as the 6th century BC the Babylonians accumulated accurate calculations of their rising and setting. The Zodiacal circle is the 'path of the sun'; its symbolic images subdivide time into qualitatively different periods. All things living in these phases, including man, share the qualities of the phases. Closer to the centre (where the earth lies) move the planets whose positions constitute the horoscope of a given moment, such as a person's birth. Time in astrology is not an empty frame of reference but has a qualitative aspect which it imparts to all things which exist in the timebound world. (Zodiac, from ms., France, c. AD 1000.)

15   Our calendar (from the periods, *calendae*, into which the Romans divided time) is based on the solar day. It is designed to keep the beginning of each year (of approximately $365\frac{1}{4}$ solar days) at the same distance from the solstices and equinoxes. In the late Roman Republic the cumulative error led to political abuses, prolonging or curtailing the terms of office of magistrates. In 46 BC Julius Caesar reformed the calendar by adding an extra day every fourth year. This still not being quite accurate, in AD 1582 Pope Gregory XIII restored the equinoxes to their proper calendrical place by suppressing ten days and ordering the extra intercalary day to be omitted in all centenary years except those which are multiples of 400. This was praised as restoring order, in place of confusion, to all human affairs. (In China the reforms of the calendar were carried out with expressly ethical intent.) According to modern calculations the Gregorian calendar is subject to an annual error of 19.45 seconds, or one day in approximately 4442 years. (Gregory XIII presides at a calendar reform meeting, anonymous painting, Archivio di Stato, Siena, Italy, 16th c.)

16, 17   Far out in the skies, man saw stellar divine images which in their timebound dance successively influenced all earthly events. They constitute an inexorable destiny (Heimarmene) which binds all temporal things. According to astrology, we are now leaving the era of Pisces and moving into a new age, that of Aquarius, which might be ruled not so much by opposites (the two fishes) as by the image of a cosmic human figure of the Self. (Planisphere with constellations of the northern and southern hemispheres arranged concentrically, from astronomical ms., France, 15th c.; Zodiac, from early medieval ms., Italy.)

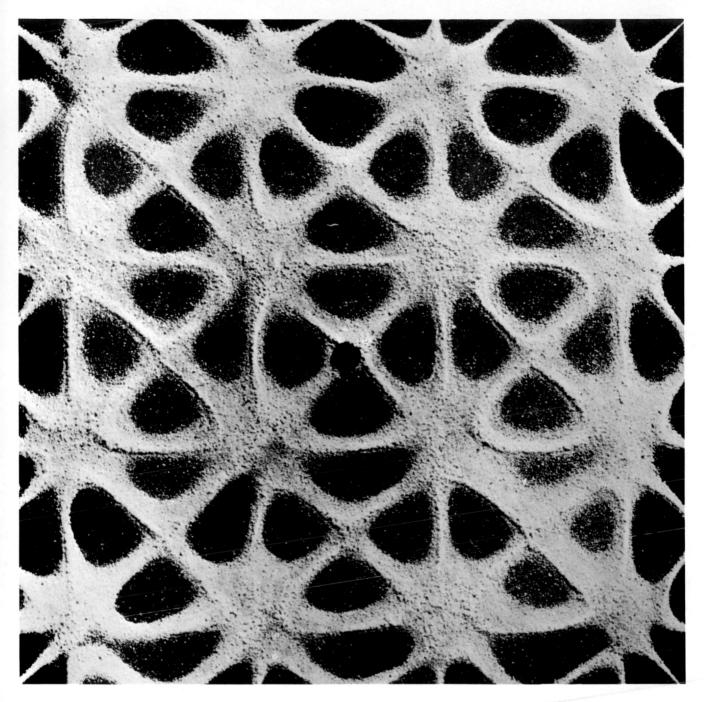

18 'Every form of life appears
to us as a *Gestalt* with a specific
development in time as well as
space. Living things, like melodies,
might be said to be configured time'
(Portmann). In flowering, a change in
the plant's inner state is manifested
in time. Plants live through phases of
wakefulness and sleep (Chandra
Bose); many of them open at or
shortly before sunrise and close at
sunset; their blossoming, wilting and
rebirth are all timebound; in the
rings of redwood treetrunks we now
count historical periods. (Unfolding
of daffodil.)

19 E.F.P. Chladni (1756–1872) discovered how one can make sound
visible by stroking with a violin bow
a metal plate sprinkled with powder.
But oscillatory, serial and periodic
phenomena are a mysterious aspect
of the whole universe, appearing in
waves, rotations, pulsations, turbu-
lences and circulations. According
to field theory, even each subatomic
particle perpetually 'sings its song',
producing rhythmic patterns of
energy. Oscillation, seriality and
periodicity make time measurable.
(Chladni 'sound figure' produced by
piezoelectric excitation.)

20–23   Animal life shows changes
of form in time. The features manifested in the final transformation to
a coloured large-winged butterfly –
prepared in the cocoon – were built
into the caterpillar during its earliest
larval periods. The specific forms of
the mature organism are prefigured
in the egg. In the larva they survive
as 'imaginal disks'. Here they lie all
ready to be set in motion in the
critical process of metamorphosis
and to unfold in final form. 'Just as in
a well-planned display of fireworks
one set piece may bear the next
latent within it, so in the life of many
insects we find in each stage a prefiguration of new organs, which
subsequently unfold in an exactly
regulated temporal process' (Portmann). (Development of African
Wanderer butterfly.)

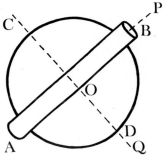

24 The astrolabe served primarily for stellar, lunar and solar altitude-taking, but one could also determine with it the points of the compass and the time. Its principle is as follows:

If a solid circle is fixed in any one position and a tube pivoted on its centre so as to move; and if the line C–D is drawn upon the circle pointing towards an object (Q) in the heavens, which lies in the plane of the circle, by turning the tube (A–B) towards any other object (P) in the plane of the circle, the angle BOD will be the angle subtended by the two objects P and Q at the eye (*Encyclopaedia Britannica*). The

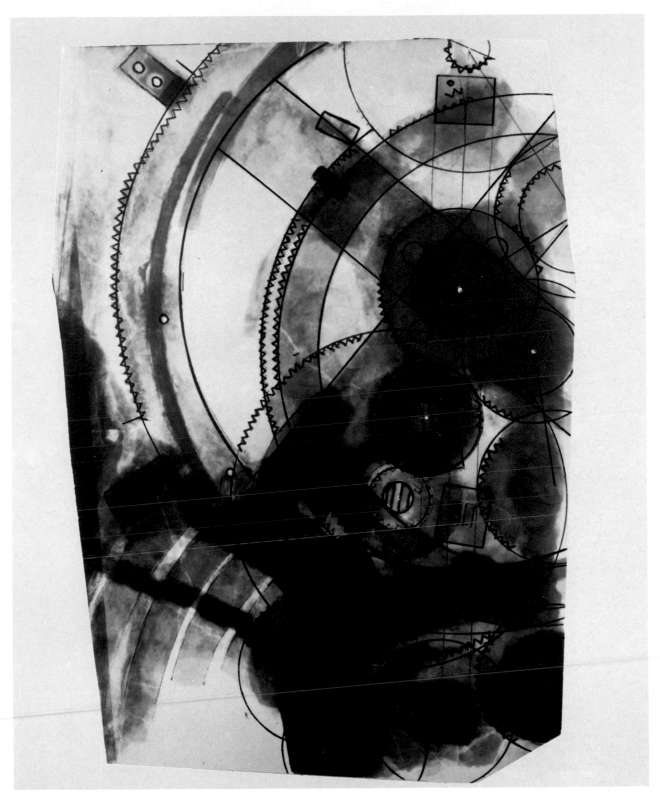

instrument above is a planispheric portable astrolabe, which was the one most widely known in the fifteenth, sixteenth and seventeenth centuries. It went out of use because it was incapable of great precision. (Astrolabe by Alfano Alfani, Italy, late 15th c.)

25   No clock could ever have been constructed before the invention of the toothed gear. Soon after 1900 a mechanism was fished out of the sea near the island of Antikythera; it showed the Zodiac and the Graeco-Egyptian calendar of twelve months of thirty days plus five epagomenic days. A pointer on two dials indi-

cated the position of sun and moon. From such antique and Islamic proto-clocks our clock has developed since the thirteenth century; it came into widespread use only at the end of the sixteenth. (Modern drawing by Professor Derek de Solla Price of instrument from Antikythera, Greece, 87 BC.)

26 Aztec basalt calendar stone carved in AD 1479. In the centre is the face of the sun god who sets the measures of time. Left and right are his claws holding human hearts, for the present Sun needs human blood to strengthen him for his journey. In the rectangular panels in the corners are the dates when the previous Suns (world aeons) were destroyed. On the top of the outer band is the date '13 Reed' when the present Sun was born. The band surrounding the claws and panels contains twenty hieroglyphs which name the days of the Aztec calendar. The next band incorporates eight V-shaped solar rays and glyphs denoting jade, or turquoise symbolizing blue, the sky and something precious. The outermost band consists of two fire serpents, which meet face to face at the bottom. They symbolize the opposites within cosmic energy. (Modern outline drawing by Roberto Sieck Flandes of Aztec Calendar Stone, Mexico, 15th c., with colouring restored.)

27 'In the night of Brahman nature is inert and cannot dance until Shiva wills it. He rises from His rapture and, dancing, sends through inert matter pulsing waves of awakening sound; and lo! matter also dances, appearing as a glory round about Him. Dancing, He sustains its manifold phenomena. In the fullness of time, still dancing, He destroys all forms and names by fire and gives new rest.' (A. K. Coomaraswamy, *The Dance of Shiva*, p. 78.) The dance of Shiva symbolizes cosmic cycles in time, of creation and destruction, birth and death. (Shiva, sculpture from the Menakshi Temple, Madras, India.)

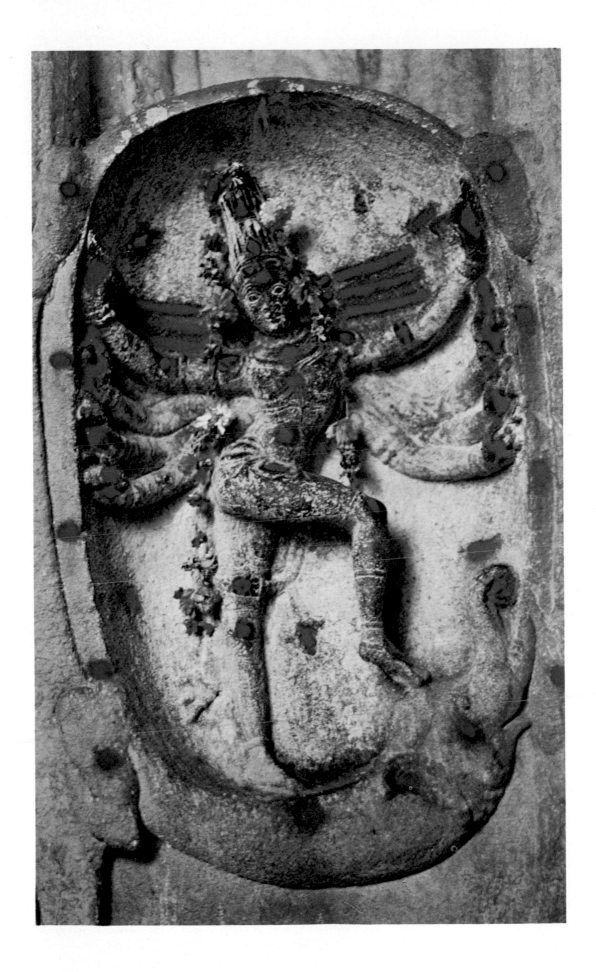

Tempus erit.

28   In antiquity the god Chronos (Time) was identified with Saturn, depicted as an old man with a sickle. In the Christian era this figure became Father Time, who is often shown devouring his children and all things. Above, he is distinct from Death: he holds the hour glass (behind him a sundial) and indicates to Death when to snuff out the light of man. He is stern but not malevolent. The inscription *Tempus erit* means 'the time will come': the time when we have to die. (Emblem by Francis Quarles, England, 1639.)

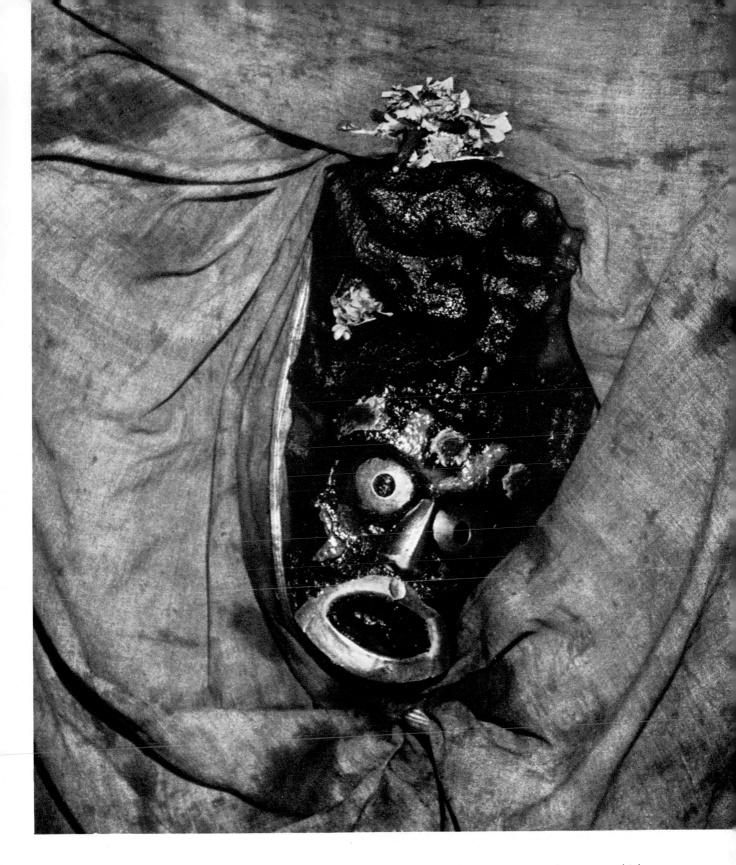

29   In the sanctuary of the Baital-
Deul temple near Aihole in India, is
this image of the goddess Durga,
who was identified with Kālī, the
female counterpart of Shiva. Kālī is
the feminine form of the Sanskrit
word Kāla (Time). 'She is black with
death and her tongue is out to lick
up the worlds, her teeth are hideous
fangs. . . . Paradoxical and grue-
some, she is today the most cher-
ished and widespread of the
personalizations of Indian cult.'
(Zimmer, 139–40.) Time in India also
meant Destiny, that power which
sets loose the effects of Karma, from
which man has to escape by an
inner awakening. (Durga,
India, c. AD 850.)

30   In Christianity, time has a beginning and an end. It is presented as a linear sequence of events through which the purpose of God moves steadily towards its goal. The end is the Last Judgment, where 'the dead, small and great, stand before God . . . . And the sea gave up the dead which were in it; and death and hell delivered up the dead which were in them' (Revelation 20:12,13). Then God creates a new heaven, a new earth, and in it the holy city of Jerusalem; there shall be no more death, and no night, but eternal light. (Last Judgment, detail from tabletop painting of *Seven Deadly Sins* by Hieronymus Bosch, Netherlands, c. 1485–1600.)

31   The all-embracing Lord Shiva Mahesvara incorporates, as Trimurti (his threefold form), simultaneously Brahman the creator, Vishnu the sustainer and Kalaruda the destroyer. He has the third eye on his forehead, his hair is braided like that of an ascetic, and he wears a snake as decoration (nāga-bhusana). Severed heads (or more often skulls) hang around his neck, for he has practised yoga on the burial grounds and is the bringer of death. The attributes in his five right hands belong to his masculine form, those in his left hands to his feminine aspect; for he is Ardhanārīsvara, the Lord Who Is Half Feminine. The triple rhythm of creation, duration and destruction repeats in immense cycles for ever and ever. (Shiva, Mandi watercolour, India, c. 1730–40.)

32 In astrophysics we meet staggering aspects of time; we can, for instance, look backward in time, to see the light of stars that have long since disappeared. The universe is in ceaseless motion; rotating clouds of hydrogen gas contract to form stars which continue to rotate, ejecting material into space. Eventually after millions of years, when most of the fuel is used up, the star expands and then contracts again into the final gravitational collapse. This collapse may turn into a black hole, where there is nothing more observable left, including time. Whether the universe has a beginning and end, or moves in cycles of expansion and collapse, or remains in a steady state in which matter is continually created anew, is still a matter of controversy. Most present-day physicists prefer the first version. (Galaxies of various types visible in Hercules.)

33   *One in all*
     All in one –
     If only this is realized
     No more worry about your not being perfect

     The believing mind is not divided
     And undivided is the believing mind
     This is where words fail
     For it is not of the past, future or present

     Seng-ts'an (Suzuki, 199–200)
     (Painting by Yasuichi Awakawa, Japan.)

34   In contrast to the
Sangsaric Wheel (12) of
Becoming or Rebirth, this is the
Wheel of Salvation. It represents the dynamic teaching
(Dharma), the Eightfold Path,
set in motion by Buddha and
rolling on and on in the world.
Gradually increasing man's
consciousness, it finally helps
him to overcome all time-
bound existence and to awake
to a timeless state of rest.
(Wheel from the Chariot
Temple of Konarak, dedicated
to the sun god Surya, India,
mid 13th c.)

Themes

The Greeks identified the god Kronos
(Saturn), who eats his own children, with
Chronos (Time). They represented him
like Saturn as a stern old man with a
scythe. Since the Middle Ages he has
become Father Time, representing a
dark aspect of the Deity. The hourglass
reminds us of the transience of all
things. (Time, statue by Ignaz Günther.
Germany, c. 1765–70. Bayerisches
Nationalmuseum, Munich.)

The stream of events

The zodiac, being the path of the sun, was represented by the Greeks as the river Oceanos which encircles the universe. It was considered to be the basic element, the world soul, from which all is engendered, even the immortal gods. Oceanos later became identified with the Persian god Zurvan (Greek: Aion), the God of 'Infinite Time' and 'Time of Long Dominion' – or creative all-embracing cosmic spirit.

The heavenly river Oceanos was also represented in Greece and Egypt as a snake. Thus on the Tomb of Seti the barge of the Sun God is driving over it. The Scarab is the agency which lifts the new-born sun over the horizon in the east every morning. Left and right of it are the eyes of Horus, all-seeing and protecting the god in the barge. (Painting from Tomb of Seti I, Thebes. Egypt, 1318–1301 BC.)

The zodiac as a tail-eating snake (Ouroboros) symbolizes the eternity of Time and the boundary of the universe. Below it the double lion Routi ('Yesterday and Tomorrow'), the agency of resurrection, supporting the new-born Sun God. (Detail of papyrus of Dama Heroub. Egypt, 21st Dynasty. Egyptian Museum, Cairo.)

In Greece it was thought that everybody had a certain life-fluid (=*aion*), which was his soul and also his allotted life-span; the Egyptians believed this too. After death the soul often appeared near the grave in the form of a snake. Here two such snake-souls move towards a sacrificial meal offered to them. (Detail of household shrine, Pompeii.)

Around the beginning of our era the Mithraic mysteries were in serious competition with Christianity to become the new world religion. Mithras, the Sun Hero, is represented slaying a bull. Spread below him is Oceanos-Kronos (Time), the dynamic Spirit of the universe, from which the Sun-hero rises. Oceanos (as Aion) was also represented as the gatekeeper of the mysteries. He is the stream of Time, the Ruler of the world, a spirit which contains all light and dark opposites. (Mithraeum of Santa Prisca, Rome.)

After Christianity won out, Oceanos still often appeared at the feet of Christ, now representing only the Ocean, father of all streams and springs. He is still a benevolent godhead but subdued by Christ, the 'New Sun' (*sol salutis*). (Oceanos. Detail of page from Sacramentary of Charles the Bald, AD 869–70. Bibliothèque Nationale, Paris.)

Measuring the flow of time

Inspired by the archetypal idea of Time as a great river, man began to measure Time by marking the flow of a substance or the creeping on of combustion. The principle is to measure time by the amount of flow from an upper receptacle to a lower one. Water clocks are, however, inaccurate because the water flows faster first and slower when there is less pressure in the upper receptacle. Before man became inflated and saw the clock as a pure mechanism, he was aware that the flowing substance and time were not of his making; therefore the hand of God is shown here pouring out the water from above. (Clepsydra. *Rudimenta mathematica*, Basle 1551.)

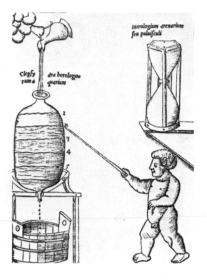

Time was not absolute, for it could be changed by miracles, interferences from eternity. Joshua could ask the sun to stand still, and Isaiah moved the sun back ten degrees by the clock for King Hezekiah during his illness (2 Kings 20:5–11 and Isaiah 38:8). (Water clock. Biblical ms., 13th c. Bodleian Library, Oxford.)

Tower of the Winds on the market of Athens, erected c. 75 BC, with nine sundials, a weathervane, a water clock and other devices. The clock incorporates a rotating disc with a star map and a model sun rotated by water power behind a wire grid representing the horizon, azimuth and altitude. The purpose was not just time-keeping, but to expound the four-element theory; this was a model of creation. (Mechanism of Tower of the Winds as reconstructed by Noble and Price. Fraser and Lawrence, II.)

Reconstruction of a water clock made by Ctesibius around 130 BC. At the base Oceanos and the Sun God. The little boys are Eros figures personifying the swift moods of love and their passing. (Reconstruction of Ctesibius' clepsydra. François Arago, *Astronomie populaire*, Paris 1857.)

A water clock from a manuscript of al-Yazari's treatise on mechanical contrivances (*Kitab fi Ma'rifat al-hiyal al handasiya*), AD 1206. At the top, the signs of the zodiac are shown; then figures successively appearing and lamps successively illuminated; below that, golden balls dropped into brazen cups from the beaks of brazen falcons to strike the chime; lastly, an automaton orchestra of five musicians (after Cresswell). Clocks were often combined with such musical devices, expressing the idea that music is, in contrast to the visual arts, the art of time configurations and rhythm. Al-Yazari also constructed a clock whose indicator was moved by the burning down of a candle which lasted 13 hours. (Water clock from al-Yazari, *On Automata*. Mamluk, AD 1206. Museum of Fine Arts, Boston, Mass.)

The oldest printed picture of a Chinese polyvascular water clock (c. 1155). On the right of the picture the author shows the oldest and simplest type of inflow clepsydra to illustrate the description. (Needham III, fig. 144.)

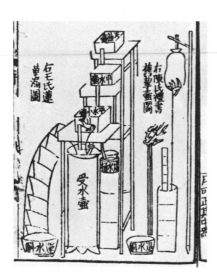

It was the Chinese who made more use of fire than of water. In their incense clock of most common use, combustible powder was spread around the labyrinthine lines; the burning of it, slowly creeping forward, indicated time. (Incense clocks. China. Science Museum, London.)

Time as an aspect or emanation of God

Kronos or Saturn (Chronos, Time) was the oldest Greek god, who lived, after being deposed by Zeus, on an island in the north, the blessed land of the dead. He was the ruler of the Golden Age. The sickle in his hand denotes that he 'harvests' everything. Macrobius praises him as the 'giver of measures'. (Chronos-Saturn. Mural from Pompeii, c. 1st c. AD. Museo Nazionale, Naples.)

In Indian thought Brahman (the Universal Spirit), or the Supreme God Vishnu, transcends Time but reveals himself also as 'Time which in progressing destroys the world'. In his sleep on the surface of the primordial ocean Vishnu 'dreams' the world; when he awakes it disappears. 4,320,000 years are one aeon; when 72×28 of them have expired, this is one day and night in the life of the god. (Vishnu reclining on the World Serpent. Dashavatara Vishnu Temple, Deogarh, India, c. AD 425.)

In Egypt the dead became one with the supreme god Atum. He is 'Yesterday, Today and Tomorrow'. The god Ḥeḥ represents 'unending Time'. He holds two measuring staves, which were the hieroglyphic sign for 'millions of years'. On his arm hangs the Ankh, symbol of life. His female counterpart Ḥeḥet was a frog-shaped goddess of birth and resurrection. (Cedar chair-back from the Tutankhamun treasure. Egypt, 14th c. BC. Cairo Museum. Ḥeḥet. After Temple of Denderah.)

Another form of the Supreme God is Shiva, who through his dance sustains the world. He is also its 'terrible destroyer', 'all-devouring Time'. His female consort Kālī (feminine of Kālā, Time) is the 'black one', for Time is irrational, hard and pitiless. She is seen treading on her husband Shiva, who placed himself as a corpse under her feet. The realization of this made Kālī cease for a moment her raving in the battlefield. (Kālī standing on Shiva. Kalighat, India, c.1880. Victoria and Albert Museum, London.)

Tonacatecutli, or Omotéotl, the supreme God of the Aztecs, is praised as 'Lord of Fire' and 'Lord of Time'. He is also the giver of good and sustains all life on earth. Through the mediation of four Tetzcatlipocas he created the universe; the four strove for dominion and thus created irreversible time. (Omotéotl. After Codex Borgia, Mexico, probably 15th c.)

The sun god as the measure of time

The cycles of the moon, and above all the course of the sun, have given man his measure of time. The sun god was worshipped as its Lord. The Sumerian god Shamash is seen here rising between two mountains at the horizon, rays of creative energy emanating from his body. At his feet (not clearly visible) are two lions who guard the eastern gates of the horizon. (Shamash scaling the twin peaks. Detail of Akkadian seal, Mesopotamia, 3rd millennium BC. British Museum, London.)

From the Djed pillar, which is either a stylized tree or the spinal cord and which symbolizes resurrection, rises the Ankh, sign of life. Above it two arms – the Ka (soul) of the sun god – and the sun disc triumphantly rising over the horizon. Baboons adore the rising sun god Ra. (Detail from Greenfield papyrus. Egypt, 21st Dynasty. British Museum, London.)

In high summer the sun is in Leo. It 'judges' all things by withering the plants which have no deep roots. The heat of the sun is thus benevolent and supports life but also brings death and illness. (Sun in Leo. Astrological ms., Cairo, Egypt, c. 1250. Bibliothèque Nationale, Paris.)

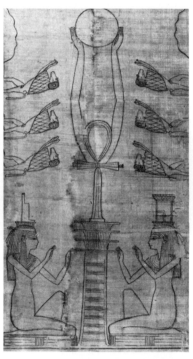

The solar hero Mithras, at the end of his work, became the son of the sun god and even identical with him. He thus brought a new order of time for mankind. From top: Oceanos and the sun god; Mithras and the sun god (twice); Mithras holding up the thigh of the slain bull. (Mithraic relief. Roman. Landesmuseum für Kärnten, Klagenfurt.)

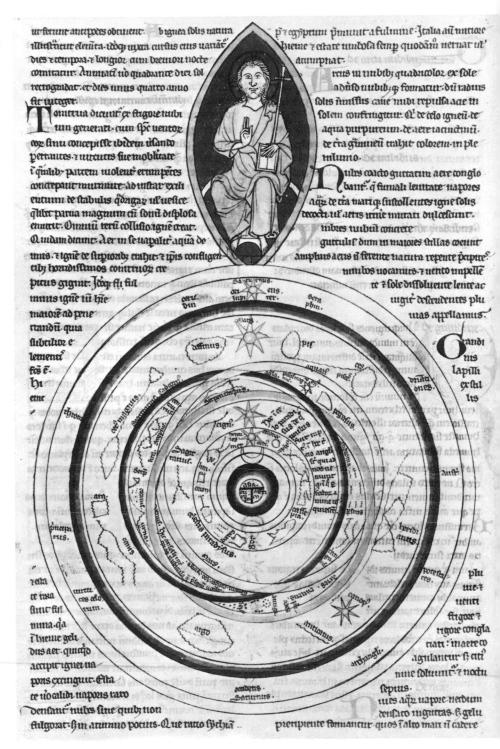

The Mexican and Maya sun god, source of cosmic energy, needs the sacrifice of human hearts to strengthen him for his journey. If he were not sustained by this, the end of the world would come sooner. (Sacrifice to sun god. Maya relief, Cotzumalhuapa, Guatemala, c. AD 1000. Staatliche Museen Preussischer Kulturbesitz, Berlin-West.)

In the Christian era Christ became the *Sol salutis* (Sun of salvation) or *Sol justitiae* (Sun of justice). The saints of the Church calendar ruled over the days; Christ ruled over the Zodiac and its influences on man. (The ordering of the seven planets, heaven and earth. Ms. of Lambert of Saint-Omer, French, c.1260. Bibliothèque Nationale, Paris.)

Measuring time by the sun

The first step in measuring time with the help of the sun is the gnomon. The place of its origin is uncertain (Chaldaea, Egypt or China). The gnomon is a rod of a known length placed vertically on an horizontal plane. The variation of the length of its shadow serves to indicate the hours of the day. (The obelisk is a more elaborate form.) The gnomon was introduced in Greece in the 6th century BC by Anaximander. Vitruvius (30 BC) mentions thirteen variations of its type. However, all gnomons are relatively inaccurate, and the more so the smaller they are. If combined with a compass which permits the rod to be placed at an angle, parallel to the axis of the earth, they can be transported from one place to another. These are called universal gnomons. If the length of the rod is varied according to the time of year, the result is an astronomical ring or gnomon. There also exist nocturnal gnomons, regulated by the course of the moon or some circumpolar star.

Two Borneo tribesmen measuring the shadow at summer solstice. (Needham, III, fig. 111.)

A late Ching representation of the solstitial measurement of the shadow by Hsi Shu in legendary antiquity, using a gnomon and a gnomon shadow template. (Needham, III, fig. 110.)

Antique sundial. (Herculaneum, c. 1st c. AD. Museo di Napoli.)

Garden sundial. (England, 1716. Science Museum, London.)

The Danish astronomer Tycho Brahe (1546–1601) accurately predicted a total solar eclipse for 21 August 1560. This was considered to be something 'divine'. He had a magnificent observatory at Uraniborg, which was later transferred to Prague under Rudolf II. (Tycho Brahe's instruments. Blaeu, *Atlas major*, 1662–65.)

Base of column sundial, 17th c. (Science Museum, London.)

Spherical astrolabe. (Islamic, by Musa, AD 1480. Museum of the History of Science, Oxford.)

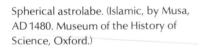

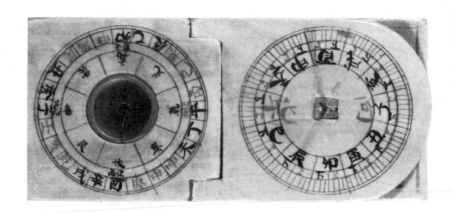

Cyclical time

In China, as in the West, there exists a cyclical conception of time. The greatest cycle is that of the Precession or migration of the spring equinox through the signs of the zodiac. Our zodiac and the Chinese one have different signs:

Ram	Rat	Scales	Horse
Bull	Ox	Scorpion	Ape
Twins	Tiger	Archer	Hen
Crab	Hare	Goat	Dog
Lion	Dragon	Water-carrier	Pig
Virgin	Snake		
		Fish	Cock

These signs were and are believed to determine the symbolic quality of specific phases in time, imprinting those events which coincide with them. (Portable sundial. China, modern. Science Museum, London. Christ with zodiac. North Italy, 11th c. Bibliothèque Nationale, Paris. Sino-Japanese zodiac. Japan. Staatliches Museum für Völkerkunde, Munich.)

The realms of existence ranging from God, angel, man, animal, plant, mineral, matter to nothingness in cyclical cosmic spheres. The uppermost and lowest are timeless and infinitely remote from the others. (After Carolus Bovillus, *De nihilo*. Mahnke, p. 110.)

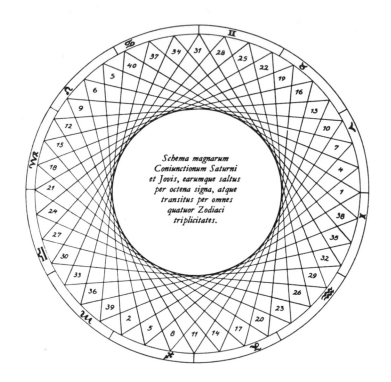

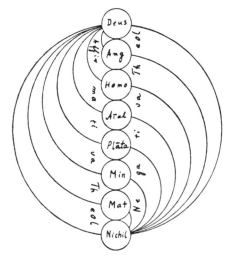

Kronos (Chronos, Saturn), being the outermost and slowest visible planet, was thought to 'give to Zeus the measures of creation'. This is Kepler's presentation of the Trigon built up by the Great Conjunctions of Saturn and Jupiter every twenty years. The motion of this Trigon along the zodiacal signs (four at a time) subdivided the cycle of the Precession. To go around the whole zodiac, one angle of the Trigon takes roughly 2400 years. (Santillana, p. 135.)

Saturn, ruler of the boundaries of the planetary circle, is the god of the melancholy, the sick, the crippled and the creative. (S. Munster, *Instrumentum planetarum*, Germany, 1522. Biblioteca Apostolica Vaticana, Rome.)

Cyclical time is forever recurring and therefore contains an element of 'eternity'. Thus man endeavoured to construct perpetual calendars which were to be valid for ever. (Perpetual calendar designed to begin in 1728 and inscribed to Isaac Newton.)

Like the days in our Church calendar, the great aeons of Indian history were marked by the lives of certain saints, which gave their aeon their spiritual imprint. The wheels probably refer to the cycles of time. (Posts at entrance to Stupa of the Saints, Sanchi. India, late 2nd c. BC.)

Similarly, astrologers have come more and more to combine mythological figures with exact numerical values. (Astronomical table. Abraham Cresque, *Catalan Atlas*, c. 1735. Bibliothèque Nationale, Paris.)

	Ienaro	Febraro	Marzo	Aprile	Mazo	Zugno	Luio	Agosto	Setébrio	Octubrio	Nouébrio	Decébrio
A												
B												
C												
D												
E												
F												
G												
H												
I												
K												
L												
M												
N												
O												
P												
Q												
R												
S												
T												

Questa sie la tauola di Salomone nelaquale se pol saper iperpetuo aquanti di del mese e aquá te boze e pori se fa la luna: e sapi che del 1490 corre p letera A: e del 1491 corre B e del 1492 corra C: e del 1491 cozzera O:ecosi ogniáno vien inginso vna letera i fino che sarai in capo cioe al T: e dapoi toz na Dacapo acomenzar dal A: e cosi ogni anno sa perai la letera che cozzer e quádo noi sapere el fare de la luna piglia la letera che corre i lano doue sei e vien p dritto fin che sei p mezo el mese doue che sei détro e la tien fermo cô el dedo che te Descbiareráno tutte le promesse e tapi che ponti uso fano vnboza. Stampata per Nicolo ditto Castilia.

The so-called Table of Solomon sets out to predict the day and time of the moon's phases for ever. The letter A gives the moons for 1490, and there is a letter for each year in a recurring nineteen-year sequence. (Lunar almanac, Venice 1488.)

The Nine Roads of the moon, a late Ching representation. The diagram shows the progressive forward motion of the major axis of the moon's orbit. The roads were traditionally assigned to symbolic colours. (Needham, III, fig. 180.)

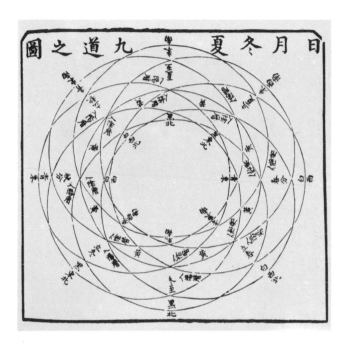

This Chinese bronze mirror, Sui to early Tang dynasty, shows in the outermost circle the signs of the Chinese zodiac. Inside, the four cardinal points represented by animals. Such mirrors were meant to represent the basic space-time structure of the universe. The four animals, generally Tiger (west), Dragon (east), Scarlet Bird (south) and Turtle, or Warrior (north), ward off misfortune and bestow longevity on the owner of the mirror. (Mirror. China, 6th c. AD. Royal Ontario Museum, Toronto.)

Time as a procession of gods

Aztec calendar stone carved at the behest of the emperor Axayácatl (1469–81). In the centre the sun god, around him panels with the dates when the previous suns were destroyed. On top of the outer band is the date '13 Reed' when the present (last) sun was born (see complete picture, pl. 26). (Calendar Stone. Mexico, 15th c. Museo Nacional de Antropología e Historia, Mexico City.)

This elaborate calendrical panel is copied from a Classic Maya monument. The top glyph at the left has no numerical value. At the right, the human figure supports a burden representing the number 144,000, which must be multiplied by the value of the human face, which is 9. In the second row, 7,200 is multiplied by 15, and 360 by 5. The third row expresses 0 times 20 and 0 times one. All these numbers added together

amount to 1,405,800. This is the number of days which must be counted forward from the beginning of the Maya calendar in the year 3113 BC. The result is a particular day in the year AD 736. All heads are necessarily in profile to allow the additions indicating the numbers, such as dots on the chin, a hand on the chin, a fleshless lower jaw or a stylized headdress. (From Stela D, Copán, Honduras. Maya, 8th c.)

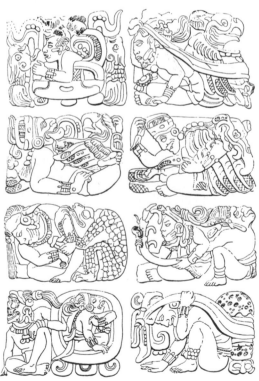

```
                        E = Spring
                5           6           2
              tiger       hare       dragon

     1   ox                                  snake   8
N   12   rat          TRADITIONAL CYCLE      horse   9    S
    11   pig                                 sheep  10

              dog        cock       monkey
               7          3           4
                        W = Autumn



                        W = Spring
                5           6           7
              tiger       hare        dog

     1   ox                                  snake   8
N   12   rat           REFORMED CYCLE        horse   9    S
    11   pig                                 sheep  10

             dragon      cock       monkey
               2          3           4
                        E = Autumn
```

In the centre four animals of eternal life: Tortoise, Kilin (Unicorn), Phoenix and Tiger, representing also the four cardinal points, are surrounded by the animals of the Chinese zodiac. Outside that are the eight Kua (world principles, basic *I Ching* signs; see pp. 25, 92). Outside them are constellations. The Chinese reformed this zodiac at some time, so that we have different arrangements: the new zodiac differs from the old only in the inversion of the sidereal position of spring and autumn. The reformed cycle is more strictly temporal than the traditional one, which was rather an eternal model of the universe.

Linear historical time

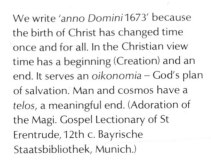

We write 'anno Domini 1673' because the birth of Christ has changed time once and for all. In the Christian view time has a beginning (Creation) and an end. It serves an oikonomia – God's plan of salvation. Man and cosmos have a telos, a meaningful end. (Adoration of the Magi. Gospel Lectionary of St Erentrude, 12th c. Bayrische Staatsbibliothek, Munich.)

The birth of Christ sheds a new light on the past as well as the future: many events in the Old Testament are understandable as prefigurations of scenes in the New Testament. When Moses raised a brazen serpent on a rod, to save the children of Israel from a snake-plague, this was a prefiguration of Christ's Crucifixion. (The Brazen Serpent. Fresco by Angelo Bronzino, Italy, 16th c. Palazzo Vecchio, Florence.)

The events of the oikonomia are both temporal and eternal. Thus, the Virgin Mary was often represented carrying in her womb God the Father and His crucified Son; for the Crucifixion was already foreseen in eternity, when she carried the child. (Or she carries in her body the scenes of her own life.) (Madonna. France, 15th c. Musée de Cluny, Paris.)

In the twelfth century, Abbot Gioacchino da Fiori divided Christian time not only into two parts, before and after the Incarnation, but into three: the age of the Father (Old Testament), the age of the Son (New Testament), and the new age of the Holy Ghost, which was to come after Gioacchino's own time, and in which the visible organization of the Church would be replaced by the contemplative monastic orders. The words of the Bible would no longer be understood literally, but symbolically, through the inspiration of the Holy Ghost. (Diagram by Gioacchino da Fiori, Italy, 12th c. Biblioteca Apostolica Vaticana, Rome.)

The Aztecs believed in a linear time. It came into existence when the four Tetzcatlipocas each strove to become the sun. Thus history is divided into five 'suns'. We are living in the last period which will end in a general cataclysm. In the middle Omotéotl, Father and Mother, Lord of Time. The four Tetzcatlipocas are on each side of the inner rectangle. The fifth reign is on the extreme right, being our own time, the last age. (The four Tetzcatlipocas. After *Codex Borgia*, Mexico, probably 15th c.)

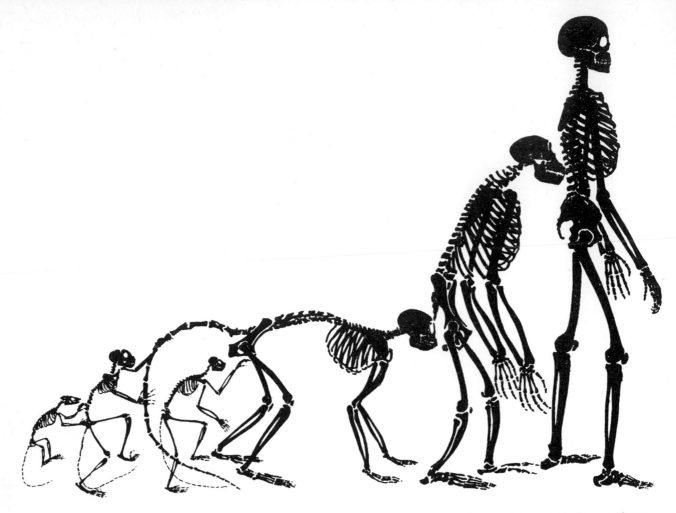

Presumed stages in the lineage of *Homo sapiens*. (W. E. Le Gros Clarke, *History of the Primates*, London 1970.)

Evolution

Over 1,100 million years have passed since traces of life have appeared on earth. All life is temporal. We can now trace the great line of evolution from plants to animals to man. Species become extinct, and mutations in the genetic code create new beings. But each living animal or plant has a *Gestalt* of its own: from seed to fully grown individual it develops step by step in a process of Becoming and Passing. The stages of man's life are the same, but also consciously experienced as a psychological change. This is the slow process of 'individuation', of becoming one's own unique self and first building up, then shedding the outer bodily frame, in order to disappear again to wherever we come from.

Metamorphosis of a butterfly. (Maria Sibylla Merian, *Metamorphosis insectorum surinamensium*, 1705.)

The Seven Ages of Man. (Engraving by Daniel Maclise, England, 19th c.)

The Ages of Man. (Popular print, Epinal, France, 1st half 19th c. Bibliothèque Nationale, Paris.)

Time as rhythm

Matter is in constant change. Each particle comes into existence, fulfils its dance, disappears again. Some live only a short flash of time, others enter into connection with others and form a combination which has a longer duration. Rhythm is a basic aspect of most forms of energy, and rhythm implies time. (Trajectories of subatomic particles in bubble chamber. CERN, Geneva.)

A most widespread rhythm of life is its reaction to the course of the sun ('circadian' rhythm, lasting one day and night). Most flowers open to the sun or shortly before it rises; animals have phases of activity and sleep in harmony with it. (Opening of an oxeye daisy, *Hencanthemum vulgare*, shot at six-hourly intervals. Specimens of the palolo worm, *Eunice viridis*, discussed on p. 21 above.)

There are not only daily (circadian) inner rhythms of life, but lunar (the human female cycle) and annual rhythms, such as the migration of birds.

	Weeks before summer solstice	Weeks after summer solstice
Golden oriole	9–6	7–9
Stonechat	17–15	15–16
Bluethroat	14–10	10–14
Stock-dove	17–14	15–17

The birds are beings 'in whom the dimension of time is extraordinarily filled with varying content, with transformations of structure and action – an extreme example of configured time' (Portmann). (Wildfowl on the Lagoon. Detail of painting by Vittore Carpaccio, Italy, 1465–c.1526. Courtesy of Christie & Co., London.)

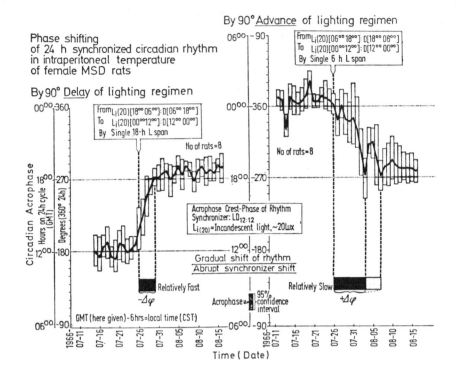

Phase shifting
of 24 h synchronized circadian rhythm
in intraperitoneal temperature
of female MSD rats

By 90° <u>Delay</u> of lighting regimen

By 90° <u>Advance</u> of lighting regimen

From L$_i$(20)[18^{00}06^{00}]·D[06^{00}18^{00}]
To   L$_i$(20)[00^{00}12^{00}]·D[12^{00}00^{00}]
By Single 18-h L span

From L$_i$(20)[06^{00}18^{00}]·D[18^{00}06^{00}]
To   L$_i$(20)[00^{00}12^{00}]·D[12^{00}00^{00}]
By Single 6 h L span

No of rats = 8

No of rats = 8

Acrophase Crest-Phase of Rhythm
Synchronizer: LD$_{12:12}$
L$_{i(20)}$ = Incandescent light, ~20 Lux

Gradual shift of rhythm
Abrupt synchronizer shift

Relatively Fast

Relatively Slow

−Δφ

+Δφ

Acrophase ⊟ 95% confidence interval

GMT (here given) − 6 hrs = local time (CST)

Time (Date)

Circadian Acrophase — Hours on 24 h cycle (GMT) — Degrees (360° 24 h)

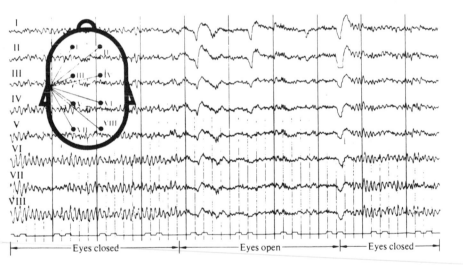

I
II
III
IV
V
VI
VII
VIII

|— Eyes closed —| |— Eyes open —| |— Eyes closed —|

Hormonal and other metabolisms are ruled by such 'biological clocks'. (Phase shifting. Fraser and Lawrence, I, 526.)

Man too has many biological clocks, rhythms which regulate metabolism, activity, etc. Our main regulator is the brain, with its various internal rhythms. Events in the cerebral cortex have a minimal duration of 70 milliseconds. Our visual sense sees a flowing process in a series of 16 to 18 images a second; our auditory sense hears tones when 18 stimuli a second reach it. These are the momentary 'impressions' of man. (Alpha, beta and other rhythms, seen on an electroencephalogram. Fraser and Lawrence, II, 65.)

Man could be called a complex living clock. In dance and music we express the rhythmicity of our whole structure – these are arts through which we relate to time and give it meaning. (The Clockmaker's Wife. Engraving by Martin Engelbrecht, 18th c.)

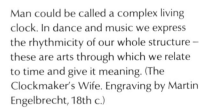

l'Horlogere.   Die Uhrmacherin.

Measuring time by rhythm

Time's rhythmicity inspired man to invent the pendulum, which was first used to drive other mechanisms than clocks. Though Leonardo da Vinci and others thought of using it for clock-making, Galileo was the first to apply a pendulum to a clock escapement. (Jacques Besson's pendulum-regulated well-bucket system, 16th c. Fraser and Lawrence, II, 422. Galileo's proposed pendulum application. Copy of drawing by Viviani, Italy, 1659. Science Museum, London.)

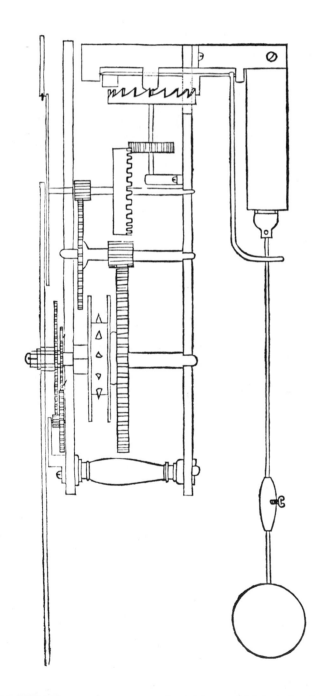

On this basis Christiaan Huygens (1626–95) developed the theory of an isochronous pendulum: if a pendulum could be made to swing on the arc of a cycloid, its period would be truly constant regardless of amplitude. (Christiaan Huygens' pendulum clock. Netherlands, c.1658. National Maritime Museum, Greenwich.)

This is a combination of sundial and clock, one dial having a moving astrolabe. In the 18th century a pendulum was added. (Gilt astrological clock. South Germany, 1550–60. British Museum, London.)

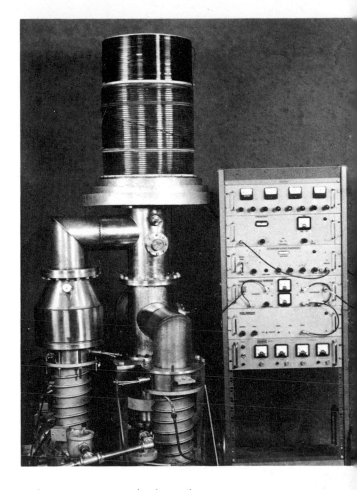

Today we measure time by the oscillations of an electrified quartz crystal, an ammonia atom, barium titanate, or other substances. Still more precise are the so-called Maser clocks (Microwave Amplification by Stimulated Emission of Radiations). For industrial purposes we often measure time by so-called 'shakes' which are 0.01 of a microsecond. Some radar instruments are accurate to one millimicrosecond, which is 0.1 shake. (Maser clock. National Physical Laboratory, Teddington.)

To the Greeks, the measures of time were the 'fetters of the universe'. What the East saw as the law of Karma was seen in the West as causality or the law of nature. At the root of the idea of causality is the Greek goddess Nemesis or Necessity, also called Dike (Justice) and Heimarmene (Astrological Fate). In quantum physics, the Heisenberg uncertainty principle has supplanted the completely determinist interpretation of natural events, but 'causality' still pre-

dominates in our scientific thinking. (Nemesis, Queen of Heaven, with Victories waiting on her; at her feet the defeated spirit of evil. Votive relief, Brindisi, Italy, early 3rd c. AD. Museo di Brindisi.)

Chance is complementary to causality. The ancient Greeks saw in Kairos (Lucky Coincidence), a winged god, whom one had to grab swiftly. Literally on a razor's edge he holds the scales whose tipping decides on fate. Often he has little wheels under his feet. (Kairos of Tragir, Dalmatia, restored. Museo di Antichità, Turin.)

In a Renaissance mural, chance is seen more as an occasion of sin. Virtue holds the young man back from grabbing his 'chance'. (Fresco of the School of Mantegna. Italy, 15th c. Palazzo Ducale, Mantua.)

Here it is not virtue but the tyranny of time which prevents man from seizing his chance. Time is Necessity, the inexorable power of man's programming mind which thinks only in terms of causality. (Engraving by G. Reverdy, 16th c.)

Whereas Fate was woven by the Moirai or Norns in paganism, the Virgin Mary, Queen of the Universe, now holds the Thread of Fate or Destiny of Mankind in her hands. (Mural painting from the church of Sant Pere de Sorpe. Spain, 12th c. Museu de Arte de Catalunya, Barcelona.)

Fortuna is the blind goddess who, with her wheel of time, carries some people upward to success and others down into misery. Here she rotates the wheel of man's lifetime. It is stopped when man rises from the grave (below: *Interruptio*). (Wheel of Fortune. From *Losbuch*, Augsburg 1461. Bayrische Staatsbibliothek, Munich.)

The numinous moment when a battle or game is won is personified in Nike (Victory), daughter of the River Styx. She is winged and makes all things obey her tune. (Lekythos by Pan Painter, Greece, c. 480–450 BC. Ashmolean Museum, Oxford.)

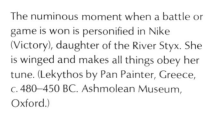

Divination

Mankind has always tried to predict events in the future by divination techniques, based on the principle of synchronicity. Most famous is the Chinese *I Ching* oracle. It is based on the permutation of unbroken (male) and broken (female) lines in eight trigrams or Kua:

☰ Heaven ☷ Earth ☲ Fire
☵ Water ☶ Mountain ☳ Thunder
☱ Lake ☴ Wind

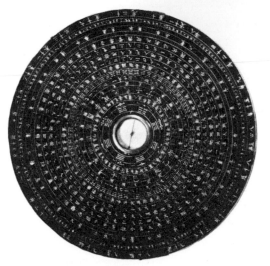

A diagram illustrating so-called 'fire-times', from the *Chou I Tschan Thung Chhi Fa Hui* (Elucidations of the Kinship of the Three and the Book of Changes). We see thirty lunar phases, five for each of six trigrams (Kua) arranged in a mirror image of the Primal Arrangement (see p. 12). In the inner circle each Kua carries its description in terms of 'nines': i.e. according to the definitions in the *I Ching*. (Needham, V, 3, 65.)

The Kua represent the basic rhythms of the universe. Compasses with these signs were also used for divination. (Chinese geomantic compass. Needham, V, fig. 338.)

Other arithmetical games were also used for divination, such as a certain form of chess and a board-game called *liu po*. Immortals playing it decided on fate; but shamans also explored the future with it. (Shamans at work. Needham, V, fig. 125.)

A Western parallel to the technique of the *I Ching* is geomancy, a combination of even or odd dots remaining if one counts off by twos a random assortment of dots. There are sixteen possibilities, which have a symbolic meaning and were combined with the partitions of the horoscope, four remaining in the centre. (European geomantic signs. Maupoil.)

Via	Populus	Cauda Draconis	Caput Draconis
o / o / o / o — 1	o o / o o / o o / o o — 2	o / o / o / o o — 9	o o / o / o / o — 10
Puer	**Puella**	**Carcer**	**Conjunctio**
o / o / o o / o — 14	o / o o / o / o — 13	o / o o / o o / o — 4	o o / o / o / o o — 3
Fortuna minor	**Fortuna major**	**Rubeus**	**Albus**
o / o / o o / o o — 5	o o / o o / o / o — 6	o o / o / o o / o o — 11	o o / o o / o / o o — 12
Tristitia	**Laetitia**	**Amissio**	**Acquisitio**
o o / o o / o o / o — 8	o / o o / o o / o o — 7	o / o o / o / o o — 15	o o / o / o o / o — 16

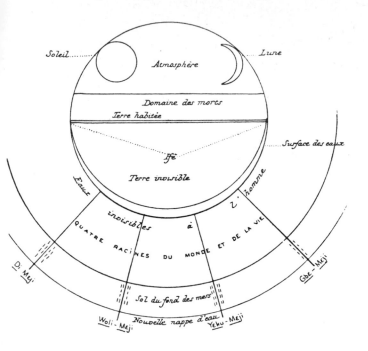

1. *Gbe-Meji*	2. *Yeku-Meji*	3. *Woli-Meji*	4. *Di-Meji*
(mâle)	(femelle)	(mâle)	(femelle)
5. *Loso-Meji*	6. *Wele-Meji*	7. *Abla-Meji*	8. *Aklā-Meji*
(mâle)	(femelle)	(mâle)	(femelle)
9. *Guda-Meji*	10. *Sa-Meji*	11. *Ka-Meji*	12. *Turukpè-Meji*
(mâle)	(femelle)	(mâle)	(femelle)
13. *Tula-Meji*	14. *Lete-Meji*	15. *Ce-Meji*	16. *Fu-Meji*
(mâle)	(femelle)	(mâle)	(femelle)

Geomancy was philosophically elaborated by the Fon in what is now the Republic of Benin in West Africa. The line combinations, called Mothers (*meji*), were interpreted as 'Roots of the universe'. (Male and female signs, Maupoil, p. 414; Life, *gbe*, as cosmic diagram, Maupoil, p. 62.)

Ce-Fu [1].

Chant.

An example of an oracle sign:

Gbo no bli baba, bo ku a,
  Ce-Fu!
Mi na bli baba, bo ku a,
  Ce-Fu!
Na wa he-nu, bo ku a,
  Ce-Fu!

The pig wallows in the mud,
  and doesn't die. Ce-Fu!
We two will wallow in the mud,
  and we won't die. Ce-Fu!
We two will do forbidden things,
  and we won't die. Ce-Fu!

This sign is an allusion to the incest between the gods Ce and Fu. It is a dangerous, unclean sign, of ill portent. (Maupoil, p. 43.)

John Horton Conway invented a 'Life Game' to simulate the rise and decline of populations. In this game a square can be occupied by a ball or not. The rules are as follows.

1 A ball in a square survives to the next phase if two or three neighbouring squares are occupied.
2 A ball is removed (death) if there are more than three or less than two balls in the neighbourhood (= overpopulation or isolation).
3 One can put a ball into a square if exactly three neighbouring squares are occupied.

Everything is mathematically determined, but the chance element lies hidden in the choice of factors represented by the balls. The scheme also fails to do justice to the much greater complexity of natural beings. In my view it is more a divinatory game than a scientific instrument. The diagram shows the fate of some configurations of triplets from Conway's Life Game. The first three sets die out after the second generation; the fourth settles into a permanent block; the fifth goes into continual oscillation. (Eigen and Winkler, pp. 218–19.)

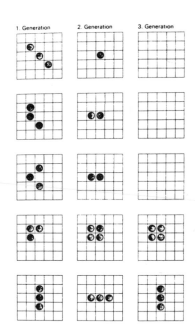

93

Transcending time

The Australian aborigines know a phase of cosmic existence called Aljira (dream time). Then, great mythological beings walked about and shaped the world. Afterwards they retired into a Beyond, but Aljira still is where the souls of the new-born come from and where the dying return. (Dream people, rock painting from Kimberley, Australia.)

In the Judaeo-Christian tradition the temporal world will come to an end. In the Book of Revelation we find the vision of the Last Judgment, the destruction of the world and the return of Christ. The subsequent 'new creation' – the heavenly Jerusalem – is outside time, eternal. (The Opening of the Sixth Seal. Beatus of Liebana, Commentary on the Apocalypse. Biblioteca de El Escorial.)

There is, with us too, a world beyond, without time. Innumerable folk tales tell how a man spends one evening in a fairy hill or subterranean sphere and discovers when he returns that over a hundred years have passed by. Rip van Winkle played at skittles with some giants one night and found himself an old man when he came back. (Scene from Washington Irving's *Rip van Winkle*, 1893 edn, illustrated by G. Boughton.)

Dante sees the City of God, while men are still suffering in this world. 'Who shall lay hold upon his mind and hold it still, that it may stand a little while, and a little while glimpse the splendour of eternity, which stands for ever: compare it with time, whose moments never stand, and see it is not comparable.' (St Augustine, *Confessions*, XI, 11.) (Allegorical portrait of Dante Alighieri, Florentine School, Italy, c.1530. National Gallery of Art, Washington, DC, Samuel Kress Collection.)

'Therefore, O monks, whatever of matter there is, whether of the past, of the future or of the present time, whether internal or external ... all matter is to be regarded as it really is, in the light of perfect knowledge, thus: "This is not of me." "This I am not. ..." Such is called one who has the obstacles removed, trenches filled, one who has destroyed, is free, one whose fight is over, who has laid down his burden, is detached.' (Majjhima Nikaia 22, Suzuki p. 150.) (Death of Buddha, detail of frieze from Gandhara, India. Smithsonian Institution, Freer Gallery of Art, Washington, DC.)

'Yoga is a cosmification of the yogi' (Eliade). His mystical body becomes a microcosm, and he annuls time by putting himself in harmony, through breathing exercises, with the great time of the universe. His inspiration then corresponds to the course of the sun, and his expiration to the moon (shown as two spiralic lines). Then he stops it altogether and concentrates his forces into the centre. Thus he transcends the temporal universe. He is a Jivan-mukta (delivered in this life) and lives no longer in time but in an 'eternal present'. (Eliade, 1962, p. 171.)

Sources and further reading

Aaronson, B. S., 'Time, Stance and Existence', *Study* I.

Aetius, *Placita Philosophorum* I, 7.22.

Anaximander, 'Simplicius in Aristotelis physicorum libros Commentarius', trs. M.-L. von Franz. Quoted from Diels, 476.

Ariotti, P. E., 'The Concept of Time in Western Antiquity', *Study* II.

Barth, 'Descartes Begründung der Erkenntnis', diss., Bern 1913.

Beauregard, Costa de, *Le Second Principe et la science du temps*, Paris 1963.

Bi-Yän-Lu, *Niederschrift von der smaragdenen Felswand*, ed. W. Gundert, Munich 1960.

Bhagavadgita, The, see Ryder.

Böhme, G., *Zeit und Zahl: Studien zur Zeittheorie bei Platon, Aristoteles, Leibniz und Kant*, Frankfurt a.M. 1974.

Bonnet, H., *Reallexikon der ägyptischen Religionsgeschichte*, Berlin 1952.

Brandon, S. G. F., 'The Deification of Time', *Study* I; *History, Time and Deity*, Manchester 1965.

Bünning, E., *Die physiologische Uhr*, 2nd edn, Berlin, New York 1963.

Campbell, L. A., *Mithraic Iconography*, New York 1968.

Capra, F., *The Tao of Physics: An Exploration of the Parallels Between Modern Physics and Eastern Mysticism*, Berkeley 1975.

Chao-Lun, *The Treatises of Seng-Chao*, trs. Walter Liebenthal, Hong Kong 1968.

Clement of Alexandria, *Stromata* IV, 17.2.

Cornford, F. M., *Plato's Cosmology*, London 1948.

Cumont, F., *Astrology and Religion among the Greeks and Romans*, New York 1912.

Da Fiori, G., *Tractatus super quattuor Evangelia*, ed. E. Bonaiuti, Rome 1930.

Dauer, D. W., 'Nietzsche and the Concept of Time' *Study* II.

De Solla Price, D. J., 'Automata and the Origins of Mechanism and Mechanistic Philosophy', *Technology and Culture* V, 1964. Quoted in De Solla Price, 'Clockwork before the Clock and Timekeepers before Timekeeping', *Study* II.

Denbigh, K. G., 'In Defence of the Direction of Time', *Study* I.

Diels, H., *Doxographi graeci*, Berlin/Leipzig 1929.

Dunne, J. W., *An Experiment with Time*, London 1964.

Eigen, M., and R. Winkler, *Das Spiel: Naturgesetze steuern den Zufall*, Munich/Zürich 1975.

Eliade, M., *The Myth of the Eternal Return*, London 1949; *Patañjali et le yoga*, Paris 1962; 'Time and Eternity in Indian Thought', in *Man and Time* (1957).

Franz, M.-L. von, 'The Dream of Descartes', *Timeless Documents of the Soul*, Evanston, Ill. 1968; *Number and Time*, Evanston, Ill., and London 1974.

Fraser, J. T., *The Voices of Time*, New York 1966; (ed.) *Of Time, Passion and Knowledge*, New York 1975.

Fraser, J. T., and N. Lawrence (eds), *The Study of Time: Proceedings of the First/Second Conference of the International Society for the Study of Time*, 2 vols, Heidelberg/New York 1972, 1975.

Gamov, G., *Atomic Energy*, Cambridge 1947.

Granet, M., *La Pensée chinoise*, Paris 1968.

Green, H. B., 'Temporal Stages in the Development of the Self', *Study* II.

Haber, C., 'The Darwinian Revolution in the Concept of Time', *Study* I.

I Ching, see Wilhelm.

Jung, C. G., *Collected Works*, ed. A. Jaffe, New York and London 1961, 1962, 1963; *Memories, Dreams, Reflections*, New York and London 1961, 1962, 1963.

Kastenbaum, R., 'Time, Death and Old Age', *Study* II.

Kausitaki Upanishad IV, 2, quoted in Eliade, *Man and Time*, 187.

Krickeberg, W., H. Trimborn, W. Müller, O. Zerries, *Pre-Columbian American Religions*, London and New York 1968.

Le Lionnais, F., *Le Temps*, ed. R. Delpire, Série Science No. 2 de l'Encyclopédie Essentielle, Paris 1959.

León-Portilla, M., *Aztec Thought and Culture*, Norman, Okla. 1963, 2nd edn 1970.

Macrobius, *Saturnalia* I, 22.

Mahnke, D., *Unendliche Sphäre und Allmittelpunkt*, Stuttgart 1937, repr. 1966.

Man and Time, papers from the *Eranos Yearbooks*, Bollingen Series xxx. 3, New York 1957.

Marshak, A., *The Roots of Civilization*, New York 1972.

Maupoil, B., *La Géomancie à l'ancienne Côte des Esclaves*, Paris 1943.

Morley, S. G., *An Introduction to the Study of the Maya Hieroglyphs*, Washington, D.C. 1915.

Müller, W., *Indianische Welterfahrung*, Stuttgart 1976. See also Krickeberg et al.

Needham, J., *Time and Eastern Man*, The Henry Myers Lecture, London 1964; Needham, J. and Wang Ling, *Science and Civilization in China*, Cambridge 1959 etc.

Onians, R. B., *The Origins of European Thought*, Cambridge 1954.

Pfeiffer, F. (ed.), *Meister Eckhart*, 1857, London 1956.

Piaget, J., *Le Développement de la notion du temps chez l'enfant*, Paris 1946.

Plato, *Timaeus*, 37 c,d, see Cornford, pp. 97–100.

Pöppel, E., 'Oscillations as a Possible Basis for Time-Perception', *Study* I.

Portmann, A., 'Time in the Life of the Organism', in *Man and Time* (1957).

Preisendanz, K., *Papyri graecae magicae*, I, II, Teubner, Stuttgart 1973.

Priestley, J. B., *Man and Time*, Aldus, London 1964.

Purce, J., *The Mystic Spiral*, London and New York 1974.

Quispel, G., 'Time and History in Patristic Literature', in *Man and Time* (1957).

Richter, C. P., 'Astronomical References in Biological Rhythms', *Study* II.

Ryder, A. W. (trs.), *The Bhagavadgita*, Chicago, n.d.

Santillana, G. de, *Hamlet's Mill*, Boston 1969, p. 135.

Saussure, L. de, *Origine babylonienne de l'astronomie chinoise*, Archives de Sciences physiques et naturelles, 5e série, 5, 198.5, 1923; *Les Origines de l'astrologie chinoise*, Maison Neuve, Paris 1930.

Schaltenbrand, G., 'Cyclic States as Biological Space-Time Fields', *Study* II.

Study, see Fraser and Lawrence (eds.).

Suzuki, D. T., *Studies in the Lankavatara Sutra*, London 1930, repr. 1957, 1968.

Thompson, J. E. S., *Maya Hieroglyph Writing*, Washington, D.C. 1950.

Usener, H., *Götternamen*, Frankfurt a.M. 1948.

Waley, A., *The Way and its Power: A Study of the Tao tê Ching*, London 1942, repr. 1949.

Watanabe, M. S., 'Causality and Time', *Study* II.

Whitrow, G. J., *The Natural Philosophy of Time*, London/Edinburgh 1961; 'Reflections on the History of the Concept of Time', *Study* I.

Whorf, B. L., *Sprache, Denken, Wirklichkeit*, Hamburg 1963.

Whyte, L. L., *Accent on Form*, New York 1954.

Wilhelm, R., *I Ching, the Book of Changes*, trs. C. Bayne, 2 vols. London 1968.

Wit, C. de, *Le Rôle et le sens du lion dans l'Egypte ancienne*, Leiden 1951.

Zimmer, H., *Myths and Symbols in Indian Art and Civilization*, New York 1962.

Acknowledgments

The objects shown in the plates, pp. 33–64, are in the collections of:
Athens, Agora Museum 11; Boulogne, Bibliothèque Municipale 14; Cairo, Egyptian Museum 8; Hamburg, Museum für Kunst und Gewerbe 24; London, British Library 3, 13, 28, Museum of Mankind 9, Victoria and Albert Museum 5, 12, 31; Madrid, Museo del Prado 30; New York, Metropolitan Museum 7; Paris, Bibliothèque Nationale 16; Rome, Biblioteca Apostolica Vaticana 17, Musei Vaticani 10; Toronto, Royal Ontario Museum 1.

Photographs and other illustrations were made available by:
PLATES Alinari 2; H. Angel 18; Archivio di Stato, Siena 15; Ardea Photographics, A. Weaving 20–23; Basilius Press 19; Roloff Beny 11, 27, 34; Bildverlag Freiburg 6; Roberto Sieck Flandes 26; Giraudon 14; Hale Observatories, Pasadena 32; Kodansha International 33; Lehnert and Landrock, Cairo 8; Pierre Rambach 29; Scala 4, 30; Professor D. de Solla Price, Yale 25; E. Tweedy 31.

THEMES Heather Angel 86 bottom l.; Alinari 70 top l., 74 bottom, 82 bottom l., 86, 90 bottom l.; British Museum (Natural History) 84 top; Cambridge University Press 74 top l. and r., 79 top r., 92 top, centre l. and r.; CERN, Geneva 86 top; J.-L. Charmet 87 bottom r.; Christie's 86 bottom r.; Deutsches Archäologisches Institut, Rome 66 bottom, 90 top r., top l.; Fototeca Unione, Rome 67 top; Gambit Inc., Boston 77 top r.; Giraudon 82 bottom r.; Greenwich, National Maritime Museum 75 bottom r., 88 bottom r.; London, British Museum (Museum of Mankind) 94 top; London, British Museum (Natural History) 84 top; London, India Office Library 70 top r.; London, Warburg Institute 90 bottom r.; R. Piper and Co., Verlag 83 bottom r.; Pierre Rambach p. 71 top l.; Ann Ronan Picture Library 68 top l., 69 top l.; Sotheby Parke Bernet and Co. 78 top; Springer Verlag 68 bottom, 87 bottom l., 88 top l.; Teddington (Middlesex), National Physical Laboratory 89 r.

Publicado por Parragon en 2013
Parragon Books Ltd
Chartist House
15-17 Trim Street
Bath BA1 1HA, Reino Unido
www.parragon.com

Texto: David Bedford
Ilustraciones: Brenna Vaughan y Henry St. Leger
Edición: Laura Baker
Diseño: Ailsa Cullen
Producción: Rob Simenton

Traducción: Míriam Torras para Delivering iBooks & Design, Barcelona
Redacción y maquetación: Delivering iBooks & Design

ISBN 978-1-4723-1802-2
Printed in China/Impreso en China

Quiero a mi abuela

PaRragon

Bath • New York • Singapore • Hong Kong • Cologne • Delhi
Melbourne • Amsterdam • Johannesburg • Shenzhen

A Ericito le encanta jugar al escondite con su abuela.
Un día, cuando la abuela lo estaba buscando…

Ericito se tapó la boca con
las patas para ahogar la risa.

—¿Pero dónde se ha metido
Ericito? —dijo la abuela—. Quiero preparar
una rica merienda y necesito que me ayude.
Ericito volvió a reír...

—Bueno —suspiró la abuela—, tendré
que preparar yo sola la merienda.
Ericito siguió de cerca a su abuela.

—Ojalá Ericito estuviera aquí para ayudarme a coger jugosas moras —dijo su abuela.

Aprovechando que su abuela no miraba, Ericito cogió **las moras más grandes** que pudo… ¡y las dejó en la cesta de su abuela!

—¡Pero cuántas moras! —se sorprendió su abuela—. Ya tengo suficientes para hacer una tarta.

Ericito corrió deprisa hacia la cocina de su abuela para encontrar el mejor escondite.

Ericito se agachó mucho para que
su abuela no lo viera.

—¡Ojalá Ericito estuviera aquí para ayudarme! —dijo su abuela.

Ericito volvió a reírse y luego se relamió cuando vio que su abuela
vertía la dulce y deliciosa miel en el bol.

—Esta miel será un regalo para Ericito —dijo su abuela.

Ericito salió sigilosamente de su escondite…

Trató de no hacer ruido mientras probaba
la miel, ¡pero estaba demasiado buena!

—¡Mmmm!
—susurró Ericito.

Y luego corrió a esconderse de nuevo. De repente...

—¡Ajá! —exclamó su abuela—.
¡Alguien ha probado mi miel! ¡Y ha
dejado huellas pegajosas!

—¡Oh, no! —Ericito todavía no quería que lo encontrara. ¡Aún no!

Su abuela siguió las
minúsculas,
diminutas
y pegajosas
huellas

que atravesaban la cocina.

—¡Alguien ha estado jugando al escondite
conmigo! —dijo sonriendo.
Y de pronto…

—¡Te he encontrado, Ericito! —exclamó su abuela.

Pero Ericito no
estaba detrás de
la mecedora.

Solo quedan unas pocas huellas pegajosas...

La abuela de Ericito siguió las huellas
que salían de la cocina hasta el jardín.

Las pegajosas huellas estaban por todos los sitios, pero parecían llegar hasta cerca de unas macetas.

—¡Esta vez sí que te he encontrado, Ericito! —exclamó su abuela.

¡Pero Ericito no estaba detrás
de las macetas! Estaba…
¡dentro de una!

¡Sorpresa!

—dijo Ericito riendo mientras
le daba un gran abrazo a su abuela.

—Oh, Ericito —le dijo su abuela—. Eres el mejor
jugando al escondite. ¡Espero que tengas hambre
porque la merienda ya está lista!

—¡Tengo hambre!
—dijo Ericito. Pero miró a su
alrededor y no vio la merienda
por ninguna parte.
—¿Dónde está? —preguntó.
Su abuela se rio.

—¡Ahora **tú**
tendrás que
encontrarla!
—le dijo.

Ericito buscó por el jardín
y pronto encontró…

galletas con miel…

y macedonia.

Después su abuela trajo una tarta
gigante de moras y miel.

¡Mmm!

—exclamó Ericito.

Era la mejor tarta que había probado nunca.

—¡Me encantan las meriendas que hace la abuela!

—gritó feliz Ericito—. ¡Yo...

quiero a mi abuela!